Uni

Springer
Berlin
Heidelberg
New York
Barcelona
Hongkong
London
Mailand
Paris
Tokio

Sasha Cyganowski
Peter Kloeden
Jerzy Ombach

From Elementary Probability to Stochastic Differential Equations with MAPLE®

Springer

Sasha Cyganowski
Tipperary Institute
Cashel Road
Clonmel Co. Tipperary
Ireland
e-mail: scyganowski@trbdi.ie

Peter Kloeden
Johann Wolfgang Goethe Universität
Frankfurt am Main
Fachbereich Mathematik
60054 Frankfurt am Main
Germany
e-mail: kloeden@math.uni-frankfurt.de

Jerzy Ombach
Jagiellonian University
Institute of Mathematics
ul. Reymonta 4
30059 Krakow
Poland
e-mail: ombach@im.uj.edu.pl

Library of Congress Cataloging-in Publication Data applied for

Cyganowski, S.:
From elementary probability to stochastic differential equations with MAPLE /
Sasha Cyganowski; Peter Kloeden; Jerzy Ombach. – Berlin; Heidelberg; New York;
Barcelona; Hong Kong; London; Milan; Paris; Tokyo: Springer, 2002
(Universitext)
ISBN 3-540-42666-3
0101 deutsche buecherei

Mathematics Subject Classification (2000): 60-01, 60-08

ISBN 3-540-42666-3 Springer-Verlag Berlin Heidelberg New York

Springer-Verlag Berlin Heidelberg New York
a member of BertelsmannSpringer Science+Business Media GmbH

http://www.springer.de

© Springer-Verlag Berlin Heidelberg 2002

Printed in Germany

The use of general descriptive names, registered names, trademarks, etc. in this publication does
not imply, even in the absence of a specific statement, that such names are exempt from the rele-
vant protective laws and regulations and therefore free for general use.

Typesetting: Camera-ready copy produced by the authors using a Springer TeX macro package
Cover design: *design & production* GmbH, Heidelberg

Printed on acid-free paper   SPIN: 10850538   46/3142db – 5 4 3 2 1 0

Preface

Measure and integration were once considered, especially by many of the more practically inclined, to be an esoteric area of abstract mathematics best left to pure mathematicians. However, it has become increasingly obvious in recent years that this area is now an indispensable, even unavoidable, language and provides a fundamental methodology for modern probability theory, stochastic analysis and their applications, especially in financial mathematics.

Our aim in writing this book is to provide a smooth and fast introduction to the language and basic results of modern probability theory and stochastic differential equations with help of the computer manipulator software package MAPLE. It is intended for advanced undergraduate students or graduates, not necessarily in mathematics, to provide an overview and intuitive background for more advanced studies as well as some practical skills in the use of MAPLE software in the context of probability and its applications.

This book is <u>not</u> a conventional mathematics book. Like such books it provides precise definitions and mathematical statements, particularly those based on measure and integration theory, but instead of mathematical proofs it uses numerous MAPLE experiments and examples to help the reader understand intuitively the ideas under discussion. The pace increases from extensive and detailed explanations in the first chapters to a more advanced presentation in the latter part of the book. The MAPLE is handled in a similar way, at first with simple commands, then some simple procedures are gardually developed and, finally, the `stochastic` package is introduced. The advanced mathematical proofs that have been omitted can be found in many mathematical textbooks and monographs on probability theory and stochastic analysis, some outstanding examples of which we have listed at the end of the book, see page 303. We see our book as supplementing such textbooks in the practice classes associated with mathematical courses. In other disciplines it could even be used as the main textbook on courses involving basic probability, stochastic processes and stochastic differential equations. We hope it will motivate such readers to further their study of mathematics and to make use of the many excellent mathematics books on these subjects.

As prerequisites we assume a familiarity with basic calculus and linear algebra, as well as with elementary ordinary differential equations and, in the final chapter, with simple numerical methods for such ODEs. Although Statistics is not systematically treated, we introduce statistical concepts such as sampling, estimators, hypothesis testing, confidence intervals, significance levels and p–values and use them in examples, particularly those involving simulations.

We have used MAPLE 5.1 throughout the book, but note that most of our procedures and examples can also be used with MAPLE 6 and MAPLE 7. All of the MAPLE presented in the book as well as the MAPLE software package stochastic can be downloaded from the internet addresses

```
http://www.math.uni-frankfurt.de/~numerik/kloeden/maplesde/
```

and

```
http://www.im.uj.edu.pl/~ombach
```

Additional information and references on MAPLE in probability and stochastics can be found there.

This book has its origins in several earlier publications of the coauthors, with Chapters 1 to 7 being an extension of similar material in the Polish language book [27] of the third author and with the remaining chapters being originating from the technical report [8] and other papers on the MAPLE software package stochastic by the first two coauthors and their coworkers as well as on the textbooks [19, 20] on the numerical solution of stochastic differential equations of the second coauthor and his coauthors.

We welcome feedback from readers.

Sasha Cyganowski, Clonmel, Ireland
Peter Kloeden, Frankfurt am Main, Germany
Jerzy Ombach, Kraków, Poland

August 31, 2001

Notation

Let us recall some standard notation and terminology. We assume that the reader is familiar with basic operations on sets, like unions, intersections and complements. In this book we will often use the operation of Cartesian product of sets. Let X and Y be given sets. Their Cartesian product is the set of all pairs having first elements in X and the second one in Y, that is:

$$X \times Y = \{(x,y) : x \in X, y \in Y\}.$$

The straightforward generalisation is as follows. Given sets $X_1, \ldots, X_n$ we define:
$$X_1 \times \ldots \times X_n = \{(x_1, \ldots, x_n) : x_i \in X_i, i = 1, \ldots, n\}.$$

If all sets X_i are the same, say $X_i = X$ for $i = 1, \ldots, n$, we often write:

$$X^n = X_1 \times \ldots \times X_n.$$

We denote by $\mathbb{N}$ the set of natural numbers, $\mathbb{Z}$ the set of integers, $\mathbb{Q}$ the set of rational numbers and $\mathbb{R}$ the set of all real numbers. We can interpret the set $\mathbb{R}$ as the straight line, the set $\mathbb{R}^2$ as the plane and $\mathbb{R}^3$ as three dimensional space. The set $\mathbb{R}^n$ will be called n-dimensional space.

A subset A of $\mathbb{R}^n$ is said to be an open set if for every point $x = (x_1, \ldots, x_n) \in A$ there exists an $\varepsilon > 0$ such that any point $y = (y_1, \ldots, y_n) \in \mathbb{R}^n$ ε-close to x, (i.e. satisfies $|x_i - y_i| < \varepsilon$ for all $i = 1, \ldots, n$), also belongs to A, $y \in A$. A subset A of $\mathbb{R}^n$ is said to be a closed set if its complement $\mathbb{R}^n \setminus A$ is open.

A function $f : \mathbb{R}^n \longrightarrow \mathbb{R}^m$ is continuous if for every point $x_0 \in \mathbb{R}^n$ and every $\varepsilon > 0$ there exists a $\delta > 0$ such that for any point x that is δ-close to x_0 its image $f(x)$ is ε-close to $f(x_0)$. Equivalently, f is continuous, if for any open set $G \subset \mathbb{R}^m$ its preimage $f^{-1}(G) = \{x \in \mathbb{R}^n : f(x) \in G\}$ is open.

Maple

As we have already mentioned above, one of the basic tools used in the book is MAPLE. It represents a family of software developed over the past fifteen years to perform algebraic and symbolic computation and, more generally, to provide its user with an integrated environment for doing mathematics on a computer. One of the most convenient features of MAPLE is that it can be operated by a user with minimal expertise with computers. Its very good "Help" system contains many examples that can be copied, modified and used immediately. On the other hand, more experienced users can take advantage of a full power of MAPLE, to write their own routines and sophisticated programs as well as to use the hundreds of already built-in procedures.

In this book we will use a number of MAPLE packages. The most useful for us is the **stats** package, which contains the routines that are most commonly used in statistics. In particular, we will be able to use built-in distributions, compute basic statistics, plot histograms and other statistical plots. Perhaps most important is that MAPLE make its possible to perform simulations such as "rolling" a die as many times as we want, see page 1. Moreover, we can simulate games and more complicated processes including Markov chains, Wiener processes, and the solutions of stochastic differential equations (SDEs) that arise as mathematical models of many real and quite complex situations. The generation and simulation of random and pseudo-random numbers will be considered in Chapter 6 of this book.

We start with simple MAPLE commands and procedures and finish with much more advanced procedures for solving, simulating and managing SDEs. While it is not essential, the reader may find it useful to consult some standard textbooks on MAPLE. The MAPLE home page

> http://www.maplesoft.com

lists over 200 MAPLE books.

Below we present some basic features of the **stats** package and use it to illustrate how to work with packages.

MAPLE ASSISTANCE 1

We load any package using **with** command.
```
>  with(stats);
```
[*anova, describe, fit, importdata, random, statevalf, statplots, transform*]

We are given names of available procedures and/or subpackages in the package. In this case all of them are subpackages. We load three of them to show corresponding procedures.

> with(describe);

[*coefficientofvariation, count, countmissing, covariance, decile,
  geometricmean, harmonicmean, kurtosis, linearcorrelation, mean,
  meandeviation, median, mode, moment, percentile, quadraticmean,
  quantile, quartile, range, skewness, standarddeviation, sumdata,
  variance*]

> with(statplots);

[*boxplot, histogram, scatterplot, xscale, xshift, xyexchange, xzexchange,
  yscale, yshift, yzexchange, zscale, zshift*]

> with(random);

[*β, binomiald, cauchy, chisquare, discreteuniform, empirical, exponential,
  fratio, γ, laplaced, logistic, lognormal, negativebinomial, normald,
  poisson, studentst, uniform, weibull*]

Now we can use any of the procedures listed above. For example we can have 30 observations from the normal distribution with mean = 20 and standard deviation = 1.

> normald[20,1](30);

21.17583957, 19.43663587, 20.23539400, 18.55744971, 18.92080377,
  19.97798535, 17.41472164, 19.55672872, 18.99671852,
  19.97213026, 21.52624486, 19.39487938, 20.16404125,
  20.65302474, 19.45894571, 22.13502584, 20.18442388,
  19.38337057, 19.55130240, 20.82382405, 20.25213303,
  20.19183012, 20.82798388, 18.25773287, 21.12307709,
  19.83943807, 18.44407097, 19.21928084, 19.48133239,
  19.74173503

We can make an appropriate histogram.

> histogram([%]);

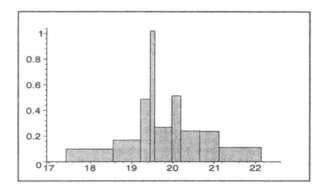

We may also want to plot the corresponding density function. We will use pdf procedure with parameter normald[20.1] from the statevalf subpackage. Still this subpackage has not been loaded. In such a case we call, for example, this procedure as statevalf[pdf,normald[20,1]].

```
>   plot(statevalf[pdf,normald[20,1]],16..24);
```

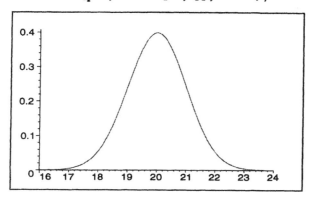

We can have the above plots at the same picture. An appropriate procedure display is contained in the plots package.

```
>   plots[display](%,%%);
```

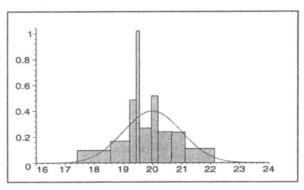

The same packages contains animate procedure.

```
>   plots[animate](statevalf[pdf,normald[20,sigma]](x),
    x = 15..25,sigma = 0.5..2);
```

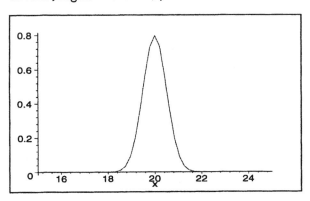

Click the plot with the mouse to see the context menu and look at the animation.

Table of Contents

1 Probability Basics

1.1 The Definition of Probability

Let us describe a simple experiment and then discuss what the word "probability" actually means.

1.1.1 Tossing Two Dice

Imagine that Mark tosses two dice and reports by internet to Tom the sum of pips obtained. Mark has been told that these numbers show some pattern and he wants Tom to find it for him. Hence, Tom is given a number from 2 through 12. After 10 trials Tom might write down, for example:

$$4, \ 7, \ 5, \ 2, \ 6, \ 11, \ 7, \ 9, \ 9, \ 6.$$

This particular sequence does not seem to show any remarkable properties. Yet, after 100 tosses Tom realises that some results occur much more often than others. Now, Tom is getting more involved in the problem. The network has been down meanwhile, unfortunately, so Tom decides to simulate the experiment using MAPLE. He writes a short program and immediately gets the following results:

MAPLE ASSISTANCE 2

```
>   die1 := rand(1..6):
    die2 := rand(1..6):
    result := NULL:
    for i from 1 to 100 do
    result := result,die1() + die2()
    od:
    result;
```

7, 10, 8, 9, 4, 6, 7, 5, 5, 10, 2, 3, 6, 4, 9, 10, 9, 7, 4, 10, 7, 7, 6, 8, 8, 7, 4, 6,
8, 8, 2, 8, 6, 5, 8, 7, 8, 9, 11, 7, 5, 7, 6, 7, 11, 8, 10, 8, 5, 3, 9, 10, 3, 9, 6,
9, 5, 6, 7, 11, 11, 6, 8, 8, 7, 7, 5, 4, 7, 10, 11, 5, 6, 7, 8, 9, 4, 8, 11, 2, 7, 12,
7, 8, 4, 4, 10, 8, 12, 6, 9, 6, 5, 9, 5, 10, 2, 7, 6, 4

Tom runs the program several times and obtains various 100 element sequences. However, all of them have the same property. Namely, the "middle" numbers 6, 7, 8 appear more often than the "extreme" numbers 2 and 12. Let us help Tom to explain the observed phenomena. We will compute the probability of the appearance of the "middle" and the "extreme" numbers. First of all, however, we should say what the word "probability" means.

The problem of defining "probability" is not easy at all. It was not resolved for more than 200 years and the solution proposed in the 1930s has caused some controversy ever since. From the mathematical point of view the matter is, however, fully explained and accepted. Such a formal definition will be much more accessible after the following informal comments are made on the preceding experiment with the dice and MAPLE.

In our experiment we were interested in the events corresponding to the values of the sum of pips on two dice. Denote the collection of all such events by Σ. There are many elements of Σ. For example, there are 11 events, say $S_2 \ldots S_{12}$, corresponding to the sum being one of the numbers from 2 to 12. There are also other events in our collection Σ. For example, the event A that the sum is 6, 7 or 8, or the event B that the sum is 2 or 12. Note that both A and B are the alternatives of the events S_6, S_7, S_8 and, respectively, S_2 and S_{12}. Also, the collection Σ contains two important, but less interesting, events; the so called impossible event (that the sum is 34, for example) and the so called certain event (that the sum is a natural number, for example).

We want to assign a number representing probability to each event from Σ. So, we have to determine a function, say P:

$$P : \Sigma \longrightarrow \mathbb{R} \quad \text{given by} \quad S \longrightarrow P(S).$$

To describe a probability, this function should satisfy a few basic properties. For example, we feel that probability is greater or equal to zero, so one property of P is that $P(S) \geq 0$ for all $S \in \Sigma$. Next, it is reasonable to have probability zero for the impossible event and probability 1 for the certain event (or 100%, if we prefer percentages over fractions). In addition, we require that the probability of the alternative of events is the sum of corresponding probabilities, assuming that the events are mutually exclusive. The latter immediately implies that: $P(A) = P(S_6) + P(S_7) + P(S_8)$ and $P(B) = P(S_2) + P(S_{12})$ for the events A and B above.

How do we compute $P(S)$ for any event S? We have to go deeper into the nature of things and, at the very beginning, answer the basic question: Why has the event S_6 occurred more often than the event S_2. The explanation is, that S_2 occurs when both dice show "1", while S_6 occurs if Mark tosses (1,5), (2,4), (3,3), (4,2) or (5,1). It is clear that S_6 occurs five times more often than S_2 does. Note that any event from the collection Σ can be identified with a set of pairs of integers ranging from 1 to 6 and corresponding to numbers on tossed dice. Denote by Ω the set of such pairs. In other words:

$$\Omega = \{(i,j) : i, j = 1, 2, \ldots, 6\}.$$

We will say that elements of Ω are elementary events. Thus any event from Σ, in fact a subset of Ω, is a collection of elementary events. It also makes sense to identify the impossible event with the empty set and the certain event with the whole set Ω. Note that elementary events may not be events themselves; the pair (2,3) is an elementary event, but it is not an event: Tom, knowing only the sum of pips, is not able to say whether or not (2,3) has occurred, while the characterising property of an event for us is that we are able to tell whether or not it has occurred. In contrast, the elementary events (1,1) and (6,6) uniquely determine events S_2 and S_{12}, and so may be thought of as events.

Each of the 36 elementary events has the same chance to occur, so it looks quite natural to assign each of them the same probability of the value $= \dfrac{1}{36}$. Now, for each $S \in \Sigma$ we put:

$$P(S) = \frac{\sharp S}{36},$$

where $\sharp S$ denotes the cardinality of the set S, that is, the number of elements in S.

Finally, we can compute $P(S_i)$ for $i = 2, \ldots, 12$ and then $P(A)$ and $P(B)$. We have:

$$P(A) = P(S_6) + P(S_7) + P(S_8) = \frac{5}{36} + \frac{6}{36} + \frac{5}{36} = \frac{16}{36}$$

and

$$P(B) = P(S_2) + P(S_{12}) = \frac{1}{36} + \frac{1}{36} = \frac{2}{36}.$$

Tom can report to Mark (Internet is alive again) that the rule is that the results "6", "7" and "8" occur 8 times more often than "2" and "12". Since Mark would not understand the whole explanation, Tom sends him a MAPLE worksheet and instructions:

MAPLE ASSISTANCE 3

Fix sets A and B:
```
> A := {6,7,8}: B := {2,12}:
```

Toss two dice 3,600 times and count how many times your result falls into A and into B:
```
> die1 := rand(1..6):
  die2 := rand(1..6):
  nA := 0: nB := 0:
  for i from 1 to 3600 do
  result := die1() + die2():
  if member(result,A) then nA := nA + 1 fi:
  if member(result,B) then nB := nB + 1 fi
  od:
```

Print the results and the fraction:

> nA,nB,nB/nA;

$$1588, \ 205, \ \frac{205}{1588}$$

Check whether the fraction is close to $\frac{1}{8}$:

> evalf(%[3] - 1/8);

.004093198992

Not satisfied with the accuracy? Toss the dice many more times, say 36,000 instead of 3,600.

1.1.2 A Standard Approach

Let us sum up the idea developed above using a slightly different point of view. The set of elementary events Ω was of central importance since with Ω we could make up the collection of events Σ in such a way that any event was a collection of elementary events, and thus essentially a subset of Ω. Moreover, for any prospective event we had to be able to tell whether it had occured or not on the basis of the experiment. Then we defined a function $P : \Sigma \longrightarrow \mathbb{R}$, which assigned to each event a number that we interpreted as the probability of this event. We also noticed that both Σ and P should obey some basic universal rules.

It appears that our approach to the problem is a rather typical one. Given a real situation with random occurrences we could essentially apply the same approach, leading to Ω, Σ and P. Certainly these objects may be dramatically different to those in the above example. For example, in some situations the set Ω will be infinite, which makes it impossible to define P in the same way as we did before. Nevertheless, in every case the collection of events Σ and the probability function P share the same universal properties.

1.1.3 Probability Space

Motivated by the above considerations we introduce the following definitions.

Definition 1.1.1 *Let Ω be a nonempty set, Σ a collection of subsets of Ω and $P : \Sigma \longrightarrow \mathbb{R}$ a function. The triple (Ω, Σ, P) is called a probability space, if the following conditions hold:*

1. *$\Omega \in \Sigma$,*
2. *If sets $A_1, A_2, A_3, \ldots \in \Sigma$, then $\bigcup_{i=1}^{\infty} A_i \in \Sigma$.*
3. *If $A, B \in \Sigma$, then $A \setminus B \in \Sigma$.*
4. *For any $A \in \Sigma$, $P(A) \geq 0$,*

5. If $A_1, A_2, A_3, \ldots \in \Sigma$ are pairwise disjoint, then

$$P(\bigcup_{i=1}^{\infty} A_i) = \sum_{i=1}^{\infty} P(A_i).$$

6. $P(\Omega) = 1$.

The elements ω of Ω are called *elementary events*, while the elements S of Σ are called *events*. Any elementary event $\omega \in \Omega$ may be considered as an event if we know that the singleton set $\{\omega\} \in \Sigma$. The set Ω represents the certain event, while the empty set $\emptyset$ represents the impossible event. The latter belongs to Σ by Condition 3 since $\emptyset = \Omega \setminus \Omega \in \Sigma$. The event $\Omega \setminus A$ is called the opposite or complementary event to A. The real number $P(A)$ is said to be the probability of A.

The above "axiomatic" definition of probability is in fact established on the basis of measure theory and has been in common use since the 1930s when it was developed by Kolmogorov. Thus, measure theory is the basic mathematical tool for studying probability. In Chapter 2 we present basic ideas, terminology and theorems about measures. At this point just note, that in measure theory a collection (set, family) Σ satisfying Conditions 1, 2 and 3 is called a *σ-algebra* and then the pair (Ω, Σ) is called a *measurable space*. A function P defined on a σ-algebra and satisfying Conditions 4, 5 and 6 is called a normalised measure or *probability measure*.

We quote and complete some properties of probability following from the corresponding properties of measures.

Proposition 1.1.1 *Let (Ω, Σ, P) be a probability space. Then:*

1. *If sets $A_1, A_2, A_3, \ldots A_N \in \Sigma$ are pairwise disjoint, then*

$$P(\bigcup_{i=1}^{N} A_i) = \sum_{i=1}^{N} P(A_i).$$

2. *For any $A_1, A_2, A_3, \ldots \in \Sigma$,*

$$P(\bigcup_{i=1}^{\infty} A_i) \leq \sum_{i=1}^{\infty} P(A_i).$$

3. *If $A \in \Sigma$, then:*
$$P(\Omega \setminus A) = 1 - P(A).$$

4. *If $A, B \in \Sigma$ and $A \subset B$, then:*

$$P(B) = P(A) + P(B \setminus A).$$

5. *If $A, B \in \Sigma$, then:*

$$P(A \cup B) = P(A) + P(B) - P(A \cap B).$$

6. *For any* $A_1 \subset A_2 \subset A_3 \subset \ldots$ *such that* $A_n \in \Sigma$ *we have:*

$$\lim_{n \to \infty} P(A_n) = P(\bigcup_{n=1}^{\infty} A_n).$$

7. *For any* $A_1 \supset A_2 \supset A_3 \supset, \ldots$ *such that* $A_n \in \Sigma$ *we have:*

$$\lim_{n \to \infty} P(A_n) = P(\bigcap_{n=1}^{\infty} A_n).$$

8. *For any* $A_1 \supset A_2 \supset A_3 \supset, \ldots$ *such that* $A_n \in \Sigma$ *and* $\bigcap_{n=1}^{\infty} A_n = \emptyset$ *we have:*

$$\lim_{n \to \infty} P(A_n) = 0$$

1.2 The Classical Scheme and Its Extensions

The best known and most commonly used probability space is the so called classical scheme.

Let Ω be a finite set of equally possible (in an intuitive sense for now) events, that is $\Omega = \{\omega_1, \ldots, \omega_N\}$, and let Σ be the family $\mathcal{P}(\Omega)$ of all the subsets of Ω. For $A \in \Sigma$, define

$$P(A) = \frac{\sharp A}{N}.$$

It is easy to see that the triple (Ω, Σ, P) is a probability space.

The classical scheme serves as a model for many simple experiments. When tossing a symmetric die we have six equally possible outcomes, so we consider Ω to be the set $\{1, 2, 3, 4, 5, 6\}$ and then the probability of any elementary event equals $\frac{1}{6}$. If we toss two symmetric dice we consider Ω to be the set of all 36 pairs of integers $1, 2, 3, 4, 5, 6$ and then the probability of any elementary event is $\frac{1}{36}$. In selecting one of 20 examination assignments, we could consider Ω to be the set $1, \ldots, 20$ with probability $\frac{1}{20}$ of any elementary event.

In these three cases we have assumed equal probability of each elementary event, which is a reasonable and natural assumption given the nature of the problems. This would not have been the case if the die in the first example was not a fair die. In such instances, the false die may produce more "1"s than "6"s. The classical scheme would not be a good model for such a situation, but we could modify it to obtain a model that is compatible with the experiment. To do this, we have to know probabilities of all the elementary events. After sufficiently long "practice" with the false die it would not be hard for an expert to determine them. We have $\Omega = \{1, 2, 3, 4, 5, 6\}$ and determine the probabilities for example as follows:

$$p_1 = 0.26; \quad p_2 = p_3 = p_4 = p_5 = 0.15; \quad p_6 = 0.14.$$

For any set $A \in \Sigma = \mathcal{P}(\Omega)$ define:

$$P(A) = \sum_{i: \omega_i \in A} p_i.$$

It is easy to see that the triple (Ω, Σ, P) is a probability space. In particular, note that we have $\sum_{i=1}^{6} p_i = 1$, which implies that $P(\Omega) = 1$.

MAPLE ASSISTANCE 4

Toss a false die for which the probabilities of each side appearing are determined as above.

```
>  die := empirical[0.26,0.15,0.15,0.15,0.15,0.14]:
>  stats[random,die](10);
          3.0, 2.0, 2.0, 3.0, 3.0, 5.0, 1.0, 5.0, 4.0, 5.0
>  results := map(trunc,[%]);
               results := [3, 2, 2, 3, 3, 5, 1, 5, 4, 5]
>  stats[random,die](1000):
>  results := map(trunc,[%]):
>  for i from 1 to 6 do
   w := x->(x=i):
   a.i := nops(select(w,results,x))
   od:
>  seq(a.i,i=1..6);
              248, 171, 151, 136, 171, 123
```

• • •

In the classical scheme the collection of events Σ is assumed to be the same as the set $\mathcal{P}(\Omega)$ for all subsets of Ω, which means that we have complete information on the experiment. It is not always the case as we noticed before, see page 1, when Tom knew the sum of the pips but not the pips themselves. In that case the class Σ is a proper subset of $\mathcal{P}(\Omega)$. In what follows we describe a general procedure of building a probability space on the basis of the classical scheme with incomplete information. This procedure generalises the method used when we helped Tom.

Define the set Ω of elementary events and the function P as we did in the classical scheme and let $S_1, \ldots S_r$ be a partition of Ω, i.e. $\bigcup_{i=1}^{r} S_i = \Omega$ and the S_i are pairwise disjoint. We define Σ to be a family of all possible unions of the sets S_i. Then Σ is a σ-algebra and the triple (Ω, Σ, P_S), where P_S denotes the restriction of P to Σ, is a probability space.

1.2.1 Drawing with and Without Replacement

We present two basic situations when the classical scheme works: drawing without replacement and drawing with replacement. Imagine that we draw n elements out of a set R of r elements and that all outcomes of this experiment are equally possible. Imagine for instance that there are $r = 10$ numbered balls in the box and we draw out $n = 5$ balls one by one. Now, there are two methods of drawing. (1) With replacement: any selected ball has its number written down and is put back into the box. (2) Without replacement: all drawn balls are left outside the box.

In case (1), an outcome is a sequence of five numbers like:

$$4, 2, 10, 5, 7; \quad 3, 3, 6, 1, 2; \quad 1, 2, 3, 4, 5 \ \text{ or } \ 7, 7, 7, 7, 5;$$

and any such a sequence represents an elementary event. Note that numbers in the sequences may repeat and that their order is essential. One has to take the order into account to make the occurence of any sequence equally like. If we disregarded the order, then the outcome $1, 2, 3, 4, 5$ would correspond to many drawings ($5! = 120$ of them) while the outcome $3, 3, 3, 3, 3$ to just one.

In case (2), an outcome is a set of five numbers like:

$$\{3, 5, 6, 7, 8\}; \quad \{1, 2, 4, 8, 10\} \ \text{ or } \ \{2, 3, 4, 5, 6\};$$

and any such set represents an elementary event. Note that in this case the numbers in the sets may not repeat and thus their order is not essential. In fact, we could take all five balls out at the same time instead of one by one.

Let us go back to the general situation and determine the sets Ω of elementary events for both methods of drawing.

For drawing with replacement, Ω is the set of all n-elements sequences taken from the r-element set R, that is Ω is the n-fold Cartesian product of the set R:

$$\Omega = R^n = \{(r_1, \ldots, r_n) : r_i \in R, \text{ for } i = 1, \ldots, n\}.$$

A simple induction argument shows that $\sharp\Omega = r^n$. For the case $n = 2$, for example, the set Ω is just the collection of all pairs of the elements of R, hence $\sharp\Omega = r^2$. So, for drawing with replacement the probability of each elementary event is

$$\frac{1}{r^n}$$

and the probability of any event A is

$$P(A) = \frac{\sharp A}{r^n}.$$

In contrast, for drawing without replacement, Ω is the set of all subsets of R containing n elements:

$$\Omega = \{N : N \subset R, \,\, \sharp N = n\}.$$

Here the cardinality of Ω is $\sharp \Omega = \binom{r}{n}$, where $\binom{r}{n} = \dfrac{r!}{n!(r-n)!}$ is a well-known combinatorial symbol. Thus for drawing without replacement the probability of each elementary event is

$$\frac{1}{\binom{r}{n}},$$

while the probability of any event A is

$$P(A) = \frac{\sharp A}{\binom{r}{n}}.$$

1.2.2 Applications of the Classical Scheme

Example 1.2.1 Suppose we mark seven numbers from 1 to 49 in a game of chance and a gambling machine chooses six winning numbers. What is the probability that we will have exactly four winning numbers.

Consider the six numbers chosen by the drawing machine. An elementary event here is any 6-element subset of the set $\{1, \ldots, 49\}$. The event A that we are interested in is the event that exactly four of "our" numbers consists of elementary events such that their four elements are chosen from the 7-element set, i.e. there are $\binom{7}{4}$ such choices, and two elements are chosen from the remaining 42-elements set $(42 = 49 - 7)$, i.e. there are $\binom{42}{2}$ such choices. Thus, the event A has cardinality $\sharp A = \binom{7}{4}\binom{42}{2}$. Hence:

$$P(A) = \frac{\sharp A}{\sharp \Omega} = \frac{\binom{7}{4}\binom{42}{2}}{\binom{49}{6}}.$$

MAPLE ASSISTANCE 5
```
>  binomial(7,4)*binomial(42,2)/binomial(49,6);
```
$$\frac{205}{95128}$$
```
>  evalf(%);
```
$$.002154991170$$

We suggest that the reader compute the same for other cases. For example, the probability of having two winning numbers is 0.1680893112. We depict this result experimentally.

MAPLE ASSISTANCE 6

Suppose we have marked the following numbers:
```
> ours := {7,23,24,28,29,36,41}:
```

We want to use `randcomb` procedure from the `combinat` package.
```
> with(combinat, randcomb);
```
$$[randcomb]$$

We simulate the gambling machine:
```
> winnings := randcomb(49,6);
```
$$winnings := \{4, 5, 9, 12, 36, 47\}$$

and see that we have only one winning number.

We will now carry out a series of 20 repetitions of the same 100 experiments. In each experiment the gambling machine draws six numbers and we check whether or not we have two of them. Then, we list the results of all series.
```
> for j from 1 to 20 do
> count := 0:
> for i from 1 to 100 do
> winnings := randcomb(49,6):
> if nops(ours intersect winnings) = 2 then count:=count+1
  fi
  od:
  clist.j := count:
> od:
  seq(clist.j,j=1..20);
```
16, 13, 11, 18, 19, 12, 21, 14, 26, 12, 16, 20, 15, 12, 19, 19, 15, 13, 19, 16

Example 1.2.2 Compute the probability that in a group of 20 students there are at least two students celebrating their birthday on the same day.

An elementary event here is 20-element sequences in which the elements correspond to the numbered days in a year. Assume for simplicity that any year is of 365 days and any day in a year is equally likely to be a birthday. Then we have a scheme of drawing with replacement. From the "box" of 365 elements we draw 20 elements out. So we have: $\#\Omega = 365^{20}$ (it is a huge number indeed). It is convenient to determine the probability of A, the complement of the event we are interested in, and then to use Property 3 in Proposition 1.1.1. The event A is the collection of all elementary events corresponding to 20-element sequences having all entries distinct. From any 20-element set one has 20! sequences and from 365 elements one has $\binom{365}{20}$ 20-element combinations, so the cardinality $\#A$ is their product and we have:

$$P(A) = \frac{20! \binom{365}{20}}{365^{20}},$$

where the probability we are seeking is actually $1 - P(A)$.

MAPLE ASSISTANCE 7

```
>  binomial(365,20)*20!/365^20;
```
$$\frac{4544397822871061676028944621910839084691488768}{772119298318740313409709163612111013793945312 5}$$
```
>  evalf(%);
```
$$.5885616164$$

So $1 - P(A) = 0.41$ (approx.). This result may look surprisingly large. Yet, we can confirm it experimentally.

MAPLE ASSISTANCE 8

First, carry out a single experiment by drawing by chance 20 days of a year:
```
>  class := NULL:
>  date := rand(1..365):
   for i from 1 to 20 do
   class := class,date()
   od:
   class;
```

2, 11, 323, 64, 222, 214, 126, 199, 352, 365, 336, 234, 188, 221, 305, 11, 277, 27, 68, 265

We see that two dates repeat (January 11), and check this result by computing the cardinality of the appropriate set.
```
>  nops({class});
```
19

Now, we carry out the series of 20 experiments and list the results of each.
```
>  sample := NULL:
   for j from 1 to 10 do
   class := NULL:
   for i from 1 to 20 do
   class := class,date()
   od:
   sample := sample,nops({class})
   od:
   sample;
```
20, 18, 19, 20, 19, 20, 19, 20, 20, 18

We see that in 5 out of 10 students group there are at least two students with their birthday on the same day.

We will explore this situation further by computing similar probabilities for various group sizes.

MAPLE ASSISTANCE 9

We insert the MAPLE spreadsheet to make the following table.

k	$\dfrac{\text{binomial}(365., k)\, k\, \Gamma(k)}{365.^k}$	$1 - \dfrac{\text{binomial}(365., k)\, k\, \Gamma(k)}{365.^k}$
5	.9728644263	.0271355737
10	.8830518223	.1169481777
20	.5885616164	.4114383836
50	.02962642042	.9703735796
100	$.3072489279\ 10^{-6}$	.9999996928

MAPLE prefers the so called Gamma function Γ to the factorial here, and so has automatically changed $k!$ into $k\,\Gamma(k)$.

1.2.3 Infinite Sequence of Events

We complete this section with an example showing that the space of elementary events may be infinite. This is essentially different to the classical scheme, but there is a common feature: each (except one) elementary event has positive probability.

Example 1.2.3 We toss a (fair) die until we throw a "6". Thus, an elementary event ω_i is that "6" appears first at the i-th tossing, where $i = 1, 2, 3, ..., \infty$. Hence

$$\Omega = \{\omega_1, \omega_2, \omega_3, \dots, \omega_\infty\},$$

where ω_∞ means we never throw a "6". Define the σ-algebra Σ by $\Sigma = \mathcal{P}(\Omega)$. Now, we try to determine the probability p_i of each elementary event ω_i. But note that the event ω_i for finite i is equivalent to drawing with replacement from the box of six balls until a specified ball will appear. It is easy to show for $i = 1, 2, 3, \dots$ that

$$p_i = \frac{5^{i-1}6}{6^i} = \left(\frac{5}{6}\right)^{i-1} \frac{1}{6}.$$

Note that $\displaystyle\sum_{i=1}^{\infty} p_i = 1$, from which we conclude that

$$p_\infty = P(\{\omega_\infty\}) = P(\Omega \setminus \{\omega_1, \omega_2, \omega_3, \ldots\}) = 1 - \sum_{i=1}^{\infty} p_i = 0.$$

For any event $A \in \Sigma$ we now define:

$$P(A) = \sum_{i: \omega_i \in A} p_i.$$

So we have constructed the probability space (Ω, Σ, P).

Now let us have a look at some elementary events in this example.

MAPLE ASSISTANCE 10

```
>   die := rand(1..6):
>   for j from 1 to 10 do
    tosses := NULL:
>   pips := 0:
>   while pips <> 6 do
    pips := die():
    tosses := tosses,pips
    od:
    print(tosses);
    od:
```

$$1, 3, 1, 1, 1, 3, 6$$
$$2, 1, 1, 6$$
$$5, 4, 6$$
$$5, 5, 5, 4, 5, 4, 6$$
$$3, 2, 6$$
$$5, 4, 2, 1, 4, 4, 5, 5, 1, 6$$
$$5, 2, 5, 4, 5, 3, 3, 3, 5, 4, 6$$
$$6$$
$$1, 5, 3, 5, 5, 2, 5, 5, 3, 2, 5, 6$$
$$6$$

Note that $\sum_{i=1}^{n} p_i = 1 - \left(\dfrac{5}{6}\right)^n$ and hence in particular: $\sum_{i=1}^{10} p_i = 0.8385$, and $\sum_{i=1}^{20} p_i = 0.974$, so there is a good chance of obtaining the first "6" in 10 tosses, while the probability of tossing a "6" in 20 tosses is close to one.

1.3 Geometric Probability

The geometric probability is applicable in cases when the elementary events can be interpreted as points in a space which has a nice geometric structure that allows us to determine length or area or volume. We use such models when we feel that, vaguely speaking, the probability of an event does not depend on its location but rather on its relative measure. We consider some examples here and will introduce the formal definition in Chapter 2, where we will specify the appropriate σ-algebras to be used.

Example 1.3.1 Irene and Mike have arranged for an internet talk between 8 p.m. and 9 p.m. If necessary, they will wait 10 minutes for each other. What is the probability that Irene and Mike will connect?

An appropriate model is to take

$$\Omega = \{(x,y) : 0 \le x \le 60, 0 \le y \le 60\}.$$

Thus, the event A that Irene and Mike connect is:

$$A = \{(x,y) \in \Omega : |x - y| \le 10\}.$$

Intuition prompts us to define the probability of A as:

$$P(A) = \frac{\mu(A)}{\mu(\Omega)} = \frac{3600 - 2\frac{50^2}{2}}{3600} = \frac{11}{36} = 0.306.$$

Here $\mu(A)$ and $\mu(\Omega)$ denote the areas of the sets A and Ω.

MAPLE ASSISTANCE 11

Simulate 100 cases for Irene and Mike and see how many times they will connect.

```
> N := 100:
> Irene := stats[random,uniform[0,60]](N):
> Mike := stats[random,uniform[0,60]](N):
> k := 0:
> for i from 1 to N do
  if abs(Irene[i] - Mike[i]) < 10 then k := k+1 fi
  od:
  k;
```
                                34

We can also graphically illustrate the situation:

```
>  restart:
>  with(plots):
>  inequal( {Irene - Mike < 10,Mike - Irene <10},
   Irene=0..60, Mike=0..60, optionsfeasible=(color=red),
   optionsexcluded=(color=green),labels=[Mike,Irene]);
```

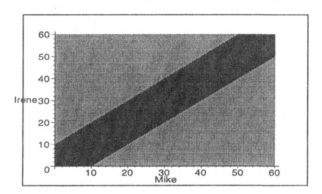

1.3.1 Awkward Corners – Bertrand's Paradox

In many cases when we have to determine a probability space, a procedure to be used is obvious. On the other hand, there are cases in which it is not obvious at all and some doubts appear. A classical one, known as Bertrand's Paradox, should be a serious warning any time we try to construct a probability space.

Example 1.3.2 What is the probability that a randomly chosen chord of the unit circle is shorter than the side of the equilateral triangle inscribed in the circle?

This problem has a number of correct solutions which lead to different results! The solution depends on the method of choosing the chord. We present two such solutions and urge the reader to look for some more.

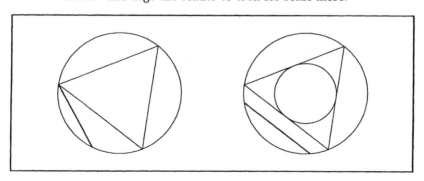

(1) Fix one end of a chord at a point on the circle and consider an equilateral triangle having one of its vertices at the same point. We choose a chord by pointing to its other end on the circle. Thus we have the unit circle as Ω and the probability of an event $A \subset \Omega$ is thus

$$P(A) = \frac{\text{length } A}{2\pi}$$

if A is any arc of the circle. We can easily compute the probability that we are looking for; it is $\frac{2}{3}$.

(2) Since a chord is uniquely determined by its mid point, we may choose such points instead of chords themselves. Here we have geometric probability with Ω being the unit disk now. A chord is shorter than the side of an equilateral triangle inscribed if its mid point is in the annulus bounded by the unit circle and the circle of radius $\frac{1}{2}$. Therefore the desired probability is $\frac{3}{4}$.

This example shows how important our interpretation of a problem may be.

1.4 Conditional and Total Probability

As a measure of uncertainty, probability depends on information. Thus, the probability we assign to the event "Microsoft stock price will go up tomorrow" depends on what we know about the company. A favourable quarterly report would definitely increase it.

The concept of the conditional probability will be introduced by the following example:

Example 1.4.1 What is the probability of at least one "6" occurring when two dice are tossed, if we know that the sum of pips is 10?

We have already known that the appropriate model for tossing two dice is the classical scheme (Ω, Σ, P) with set of elementary events $\Omega = \{(i,j) : i,j = 1, \ldots 6\}$. Denote by B the set of those pairs $(i,j) \in \Omega$ with either i or j equal to 6 and by the set A those pairs $(i,j) \in \Omega$ with $i + j = 10$. One can easily check that $\sharp B = 11$, while $\sharp A = 3$. What we want to do is to compute the probability of B if we know that A has just occurred. Now the set of all elementary events is no longer the whole set Ω, but rather the set A. Moreover, we are no longer interested in the whole set B, but rather in its intersection with the set A, that is $A \cap B$, which has just two elements. So the probability we want is $\frac{\sharp(A \cap B)}{\sharp A} = \frac{2}{3}$.

We can obtain the same result extending the idea of geometric probability established above. Namely, in this interpretation, it makes sense to determine the required probability as:

$$\frac{P(A \cap B)}{P(A)},$$

and so to obtain: $\dfrac{2}{36} \Big/ \dfrac{3}{36} = \dfrac{2}{3}$.

We are ready to introduce the following definition.

Definition 1.4.1 Let (Ω, Σ, P) be a probability space and let $A \in \Sigma$ be such that $P(A) > 0$. Then for each $B \in \Sigma$ we define

$$P(B|A) = \frac{P(A \cap B)}{P(A)} \tag{1.1}$$

and call it the conditional probability of B given A.

1.4.1 The Law of Total Probability

There is a simple relationship between conditional probabilities and the "total" probability of an event (a set). This is known as the Law of Total Probability.

Theorem 1.4.1 Let (Ω, Σ, P) be a probability space and let the events $A_1, \ldots, A_n \in \Sigma$ satisfy:

(i) $P(A_i) > 0$ for each $i = 1, \ldots, n$,
(ii) $A_i \cap A_j = \emptyset$ for all $i \neq j$,
(iii) $A_1 \cup \ldots \cup A_n = \Omega$.

Then, for any event $B \in \Sigma$ we have

$$P(B) = \sum_{i=1}^{n} P(B|A_i)P(A_i), \tag{1.2}$$

i.e. given a collection of mutually exclusive conditioning events whose union is the certain event, we can compute the probability of any event provided we know all the conditional probabilities of this event given the conditioning events and the probabilities of the conditioning events themselves.

Proof. Since

$$B = B \cap \Omega = B \cap \left(\bigcup_{i=1}^{n} A_i\right) = \bigcup_{i=1}^{n} (B \cap A_i),$$

and $B \cap A_i$ are pairwise disjoint we have

$$P(B) = \sum_{i=1}^{n} P(B \cap A_i) = \sum_{i=1}^{n} P(B|A_i)P(A_i).$$

□

The Law of Total Probability is a powerful tool. One of its big advantages is that sometimes one can compute probabilities without having to specify a probability space.

Example 1.4.2 In a "One Question Competition" students are required to answer a single question from a known list of 200 problems in biology, 100 problems in chemistry and 100 problems in physics. A student believes he can answer 150 of the problems in biology, all of the problems in chemistry and 80 of the problems in physics. What is the probability that he will be able to answer the single question that is asked?

Solution 1. We have the certain union of mutually exclusive conditions, say B, C, P, that a problem we are asked for is from biology, chemistry and physics, respectively. We want to compute the probability of the event A that we will answer correctly for a single problem. We know from the contents:

$$P(B) = \frac{200}{400} = \frac{1}{2} \quad P(C) = \frac{100}{400} = \frac{1}{4} \quad P(P) = \frac{100}{400} = \frac{1}{4}$$

$$P(A|B) = \frac{150}{200} = \frac{3}{4} \; P(A|C) = \frac{100}{100} = 1 \; P(A|P) = \frac{80}{100} = \frac{4}{5}$$

From the Law of Total Probability we have:

$$P(A) = \frac{3}{4} \cdot \frac{1}{2} + 1 \cdot \frac{1}{4} + \frac{4}{5} \cdot \frac{1}{4} = \frac{33}{40} = 0.825.$$

There is a 82.5% chance to answer the question.

Solution 2. We consider 400 problems and we can answer 330 of them. So, using the classical scheme with $\Omega = \{1, \dots, 400\}$ we immediately have:
$$P(A) = \frac{330}{400} = 0.825.$$

In the following example there is no obvious way to construct the probability space.

Example 1.4.3 A drug company estimates a chance of 90% that the government will approve a new drug if the results of current testing show no side effects of the drug. Further, it estimates a chance of 40% of approval if the tests show some side effects. Physicians working for the company believe there is a 0.1 probability that tests will show side effects of the drug. What is the probability that the drug will be approved?

We have the certain union of mutually exclusive conditions EY and EN that the tests will show side effects or will not show side effects, respectively. Denote by A the event that the drug will be approved. Then:

$$P(EY) = 0.1 , \quad P(EN) = 0.9$$

$$P(A|EY) = 0.4 , P(A|EN) = 0.9$$

and hence: $P(A) = 0.4 \cdot 0.1 + 0.9 \cdot 0.9 = 0.85$.

MAPLE ASSISTANCE 12

We can check the above results experimentally. First, write down what we actually know about the problem and print ten possible results of the tests for a try:

```
>   sideefect := empirical[0.1,0.9]:
    approval_Yes := empirical[0.4,0.6]:
    approval_No := empirical[0.9,0.1]:
    stats[random,sideefect](10);
            2.0, 2.0, 2.0, 1.0, 2.0, 2.0, 2.0, 2.0, 2.0, 2.0
```

Here 1.0 means side effects disclosed and 2.0 no side effects. Let us make it more readable.

```
>   F := proc(x)
    if x = 1.0 then YES elif x = 2.0 then NO
    fi
    end:
>   results := map(F,[%%]);
    results := [NO, NO, NO, YES, NO, NO, NO, NO, NO, NO]
```

Make $N = 15$ simulations of our case.

```
>   N := 15:
>   stats[random,sideefect](N):
    results := map(F,[%]);
```

$results := [NO, NO, NO, NO, NO, NO, NO, NO, YES, NO, NO, NO,$
$YES, YES, NO]$

```
>   for i from 1 to N do
    if results[i] = YES then
    decision[i] := stats[random,approval_Yes](1) else
    decision[i] := stats[random,approval_No](1)
    fi
    od:
>   approval := seq(F(decision[i]),i=1..N);
```

$approval := YES, YES, YES, NO, YES, YES, YES, YES, YES, YES,$
YES, YES, YES, NO, YES

How many approvals do we have?

```
>   w := x->x=YES:
>   nops(select(w,[approval],x));
```

13

The fraction of approvals is then:

```
>  %/N;
```

$$\frac{13}{15}$$

Let us carry out many more simulations, first modifying the above command lines to avoid unwanted printouts.

```
>  N := 1500:
>  stats[random,sideefect](N):
   results := map(F,[%]):
>  for i from 1 to N do
   if results[i] = YES then
   decision[i] := stats[random,approval_Yes](1) else
   decision[i] := stats[random,approval_No](1)
   fi
   od:
>  approval := seq(F(decision[i]),i=1..N):
>  nops(select(w,[approval],x));
```

$$1272$$

```
>  evalf(%/N);
```

$$.8480000000$$

$$\cdots$$

Remark 1.4.1 *One can similarly establish the Law of Total Probability for a countable collection of conditions* $A_1, A_2, A_3, \ldots$.

1.4.2 Bayes' Theorem

The Law of Total Probability implies a theorem established by an English clergyman Bayes in the eighteenth century. Despite its simplicity Bayes' Theorem has become entrenched in mathematics and continues to give rise to intensive debate on the meaning and usage of probability. Generally speaking, it says that one can estimate probabilities of causes on the basis of their effects.

Theorem 1.4.2 *Consider the same assumptions as in the Law of Total Probability, Theorem 1.4.1, with additionally,* $P(B) > 0$. *Then, for any* $k = 1, \ldots, n$ *we have:*

$$P(A_k|B) = \frac{P(B|A_k)P(A_k)}{\sum_{i=1}^{n} P(B|A_i)P(A_i)}. \tag{1.3}$$

Proof. From the definition of conditional probability, applied twice, we have:

$$P(A_k|B) = \frac{P(A_k \cap B)}{P(B)} = \frac{P(B|A_k)P(A_k)}{P(B)}.$$

Applying Theorem 1.4.1 completes the proof.   □

Example 1.4.4 Experience gained over a few last years shows that the written part, 60% pass, of the examination on probability theory is much harder than the oral part, 95% pass, of this examination. To pass the whole examination one has to pass both of its parts and the rule is that one must pass the written part first to be allowed to take the oral one. What is the probability that a student who fails the examination, has failed the written part?

The conditions here are P_w and F_w that a student passed or failed, respectively, the written part. From the context we know their probabilities, namely:

$$P(P_w) = 0.6, \qquad P(F_w) = 0.4.$$

We want to compute $P(F_w|F)$, where the event F means the failing the whole examination. Actually, we also know the conditional probabilities:

$$P(F|P_w) = 0.05, \quad P(F|F_w) = 1.$$

Hence, from the Bayes' Theorem we have:

$$P(F_w|F) = \frac{P(F|F_w)P(F_w)}{P(F|P_w)P(P_w) + P(F|F_w)P(F_w)} = \frac{1 \times 0.4}{0.05 \times 0.6 + 1 \times 0.4} = 0.93.$$

MAPLE ASSISTANCE 13

We can simulate the above case to confirm the result. As in the last MAPLE Assistance we write down our data in an appropriate form.

```
>   written := empirical[0.6,0.4]:
    oral_p := empirical[0.95,0.05]:
    oral_f := empirical[0.0,1.0]:
```

and introduce an auxiliary procedure for the conversion purpose.

```
>   F := proc(x)
    if x = 1.0 then PASS elif x = 2.0 then FAIL
    fi
    end:
```

We run 1000 simulations: Imagine that one thousand students want to pass this unpleasant examination.

```
>   N := 1000:
>   stats[random,written](N):
    res_of_w := map(F,[%]):
>   for i from 1 to N do
    if res_of_w[i] = PASS then
```

```
    oral[i] := stats[random,oral_p](1) else
    oral[i] := stats[random,oral_f](1)
    fi
    od:
>   exam := seq(F(oral[i]),i=1..N):
>   w := x->x=FAIL:
>   n_of_fail := nops(select(w,[exam],x));
                    n_of_fail := 419
```

How many students failed both the written part and the whole examination?

```
>   count := 0:
    for i from 1 to N do
    if exam[i] = FAIL and res_of_w[i] = FAIL
    then count := count +1
    fi
    od:
>   count;
                    397
```

Compute the fraction:

```
>   evalf(count/n_of_fail);
                    .9474940334
```

. . .

In defining conditional probability we have assumed that the probability of the condition A is positive, an assumption which made it possible to divide by $P(A)$ in (1.1). It is possible to generalise the definition to cover some cases in which probabilities of conditions are zero and we will discuss them in Section 7.1. Now, we mention just a typical example here.

Example 1.4.5 We choose at random a number x from the unit interval $[0, 1]$ and then a number y from the interval $(0, x)$. What is the probability $P(y < \frac{1}{3} | x = \frac{1}{2})$? We believe intuitively it is $\frac{1}{3} / \frac{1}{2} = \frac{2}{3}$, yet formally it is undefined since $P(a = \frac{1}{2}) = 0$.

1.5 Independent Events

The concept of independence is one of the most fundamental in probability theory and statistics. It is this concept that distinguishes probability theory from measure theory.

Let us begin from a pair of events. We can agree to say that events A and B are independent if the conditional probability $P(B|A)$ does not depend on A and $P(A|B)$ does not depend on B. Assuming, for instance, that $P(A) > 0$, then we have from (1.1):

$$P(A)P(B) = P(A \cap B).$$

This formula makes sense whether or not $P(A)$ and $P(B)$ vanish and its generalisation will be used in the following formal definition of independence.

Definition 1.5.1 Let (Ω, Σ, P) be a probability space and let events $A_1, \ldots, A_n \in \Sigma$. The events $A_1, \ldots, A_n \in \Sigma$ are independent, if

$$P(A_{k_1} \cap \ldots \cap A_{k_r}) = P(A_{k_1}) \cdot \ldots \cdot P(A_{k_r})$$

for any integers $1 \leq k_1 < \ldots < k_r \leq n$, $1 < r \leq n$.

Example 1.5.1 We now give some examples of independent events and dependent events. Imagine we are tossing two dice. Let A denote "6" on the first die, B denote an odd number on the other and let S denote the sum of pips equals 10. Intuition prompts us to recognise: (1) A and B to be independent: the numbers on the first die have nothing to do with the numbers on the other. (2) A and S to be dependent: "6" on the first die would increase the chance for the sum to equal 10. (3) B, S to be dependent: it is not so obvious, still the information about an odd number appearing on the second die decreases the chance for the sum being 10.

Now we shall give formal arguments for the above naive statements. We use the classical scheme here with the space of elementary events $\Omega = \{(\omega_1, \omega_2) : \omega_1, \omega_2 = 1, \ldots, 6\}$. It is easy to find the cardinalities and then the probabilities that the events have:

case (1): $P(A \cap B) = \dfrac{3}{36} = \dfrac{1}{12}$, $P(A)P(B) = \dfrac{1}{6} \cdot \dfrac{1}{2} = \dfrac{1}{12}$.

case (2): $P(A \cap S) = \dfrac{1}{36}$, $P(A)P(S) = \dfrac{1}{6} \cdot \dfrac{3}{36} = \dfrac{1}{72}$.

case (3): $P(B \cap S) = \dfrac{1}{36}$, $P(B)P(S) = \dfrac{1}{2} \cdot \dfrac{3}{36} = \dfrac{1}{24}$.

which confirms our earlier observations.

The above example is somewhat artificial. In the vast majority of practical problems we do not check the independence by the definition, but we have to decide instead whether some events are independent, and if they really are, then we apply the definition or rather theorems and formulas based on the assumption on independence.

Example 1.5.2 The powering of a lift is provided by two independent systems with known reliabilities of 99% and 95% respectively. What is the probability that the lift will operate safely?

Let A_1 and A_2 denote events that the lift is powered by the corresponding system. We want to find the probability of the union, i.e. $P(A_1 \cup A_2)$. By Property 5 in Proposition 1.1.1 we have:

$$P(A_1 \cup A_2) = P(A_1) + P(A_2) - P(A_1 \cap A_2).$$

Assuming that the events A_1 and A_2 are independent, from the definition of independence we obtain:

$$P(A_1 \cup A_2) = P(A_1) + P(A_2) - P(A_1)P(A_2) = 0.99 + 0.95 - 0.99 \times 0.95 = 0.9995.$$

1.5.1 Cartesian Product

The notion of independence is closely related to the concept of Cartesian product of probability spaces. First, we discuss the simplest case of two spaces, and then briefly mention its generalisation.

Assume that $(\Omega_1, \Sigma_1, P_1)$ and $(\Omega_2, \Sigma_2, P_2)$ are two probability spaces. Define $\Omega = \Omega_1 \times \Omega_2$, the Cartesian product of sets Ω_1 and Ω_2, that is

$$\Omega = \{(\omega_1, \omega_2) : \omega_1 \in \Omega_1, \omega_2 \in O_2\}.$$

One can construct a σ-algebra Σ on the set Ω and a probability measure $P :$ $\Omega \longrightarrow \mathbb{R}$ (it is a rather sophisticated procedure[1]) such that for any $A_1 \in \Sigma_1$, $A_2 \in \Sigma_2$, we have inclusion $A_1 \times A_2 \in \Sigma$ and:

$$P(A_1 \times A_2) = P_1(A_1)P_2(A_2).$$

Sometimes we simply write: $P = P_1 \times P_2$.

Example 1.5.3 Let $\Omega_1 = \Omega_2 = [0, 1]$. The set $\Omega = \Omega_1 \times \Omega_2$ is the unit square. For any segments $A_1 \subset \Omega_1$ and $A_2 \subset \Omega_2$ their Cartesian product is a rectangle with the area equal to the product of the segments lengths. Thus the above formula is satisfied if we consider P_1 and P_2 as the lengths and P as the area. More precisely, one can prove that the Cartesian product (Ω, Σ, P) of two copies of the probability space $([0, 1], \mathcal{B}([0, 1]), \mu_1|_{[0,1]})$ is exactly the same as $([0, 1]^2, \mathcal{B}([0, 1]^2), \mu_2|_{[0,1]^2})$, where $\mu_1|_{[0,1]}$ is the 1-dimensional Lebesgue measure restricted to the interval $[0, 1]$ and $\mu_2|_{[0,1]^2}$ is the 2-dimensional Lebesgue measure restricted to the square $[0, 1]^2$.

As another example of the Cartesian product consider the probability space modeling the tossing of two dice discussed before.

Example 1.5.4 We consider two copies of the classical schemes with the space of elementary events

$$\Omega_1 = \Omega_2 = \{1, 2, 3, 4, 5, 6\}.$$

The Cartesian product is also a classical scheme with the space of elementary events

$$\Omega = \{(\omega_1, \omega_2) : \omega_1, \omega_2 = 1, \ldots, 6\}.$$

---

[1] One takes Σ to be the smallest σ-algebra containing the products of all $A_1 \in \Sigma_1$, $A_2 \in \Sigma_2$ and tediously proves existence of a unique measure P satisfying the required conditions.

The relationship between the independence of events and the Cartesian products is transparent in the following example.

Example 1.5.5 Let $(\Omega_1, \Sigma_1, P_1)$ and $(\Omega_2, \Sigma_2, P_2)$ be probability spaces and let the events $A_1 \in \Sigma_1$, $A_2 \in \Sigma_2$ be fixed. Then, the events $A = A_1 \times \Omega_2$, $B = \Omega_1 \times A_2$ are independent in the Cartesian product (Ω, Σ, P) of these spaces. This follows from the definition of the measure $P = P_1 \times P_2$ since:

$$P(A \cap B) = P(A_1 \times A_2) = P_1(A_1) \cdot P_2(A_2) = P(A) \cdot P(B).$$

$$\cdots$$

The concept of the Cartesian product and the above examples and remarks generalise straightforwardly to the case of finite many probability spaces. One can establish the definition by induction as well as a definition which repeats that one for two probability spaces; they lead to the same object.

1.5.2 The Bernoulli Scheme

The concept of independence plays an essential role in a construction of the Bernoulli scheme.

Imagine that the same experiment or trial is repeated a number of times under the following conditions.

1. Each experiment has exactly two possible mutually exclusive outcomes called success and failure and denoted by 1 and 0, respectively.
2. The probability of success p remains constant from experiment to experiment. Hence $q = 1 - p$ is the probability of failure in each experiment.
3. The experiments are independent, i.e. the outcome of any one experiment does not affect the outcomes of the other experiments.

The above conditions are satisfied, for example, when one tosses a die. Certainly, one has to decide what success and failure mean in these trials. So, let success mean "6" pips. We then have $p = \frac{1}{6}$ and $q = \frac{5}{6}$. A more general example when the above conditions are satisfied is that of drawing with replacement.

We establish a probability space modeling a finite sequence of experiments satisfying the above assumptions.

To the i-th experiment corresponds simple probability space: $(\Omega_i, \Sigma_i, P_i)$, where $\Omega_i = \{0, 1\}$, Σ_i is the σ-algebra of the all subsets of Ω_i (there are exactly four such subsets), and the measure P_i is uniquely determined by:

$$P_i(\{1\}) = p, \quad P_i(\{0\}) = 1 - p, \quad i = 1, \dots, n.$$

The independence of experiments (in the above intuitive sense) and what we have already said about the Cartesian product suggest that the probability space could be the n-th Cartesian product of $(\Omega_i, \Sigma_i, P_i)$. Define $(\Omega, \mathcal{P}(\Omega), P)$ to be such a product[2]:

$$\Omega = \{\omega = (\omega_1, \ldots, \omega_n) : \omega_i \in \Omega_i\},$$

$$P(\omega_1, \ldots \omega_n) = P_1(\omega_1) \cdot \ldots \cdot P(\omega_n)$$

and then for any subset $A \subset \Omega$:

$$P(A) = \sum_{\omega : \omega \in A} P(\omega).$$

This probability space is called a (finite) Bernoulli scheme.

In the following example we solve a classical problem to find the probability of k successes in a finite Bernoulli scheme.

Example 1.5.6 What is the probability of exactly k successes during n experiments satisfying the above Conditions 1, 2, 3?

Denote A to be the event we are interested in. Thus A consists of sequences $\omega = \{\omega_1, \ldots, \omega_n\}$ having k ones and $n-k$ zeros as the entries. For any such ω we have: $P(\omega) = p^k(1-p)^{n-k}$. So:

$$P(A) = \sum_{\omega \in A} P(\omega) = \sum_{\omega \in A} p^k(1-p)^{n-k}.$$

It suffices to find the cardinality of A, that is just the number of all k-element subsets chosen from the n-th element set. Such a subset determines k standings for ones out of n places. Hence:

$$P(A) = \binom{n}{k} p^k(1-p)^{n-k}. \tag{1.4}$$

MAPLE ASSISTANCE 14

We will simulate a sequence of Bernoulli trials in which a probability of success is $p = 0.2$.

```
>  F := proc(x)
   if x = 1.0 then 0 elif x = 2.0 then 1 fi
   end:
>  p := 0.2:
>  trial := empirical[1-p,p]:
>  stats[random,trial](10):
   outcomes := map(F, [%]);
             outcomes := [1, 0, 1, 0, 0, 0, 0, 0, 1, 0]
```

---

[2] To improve readability we omit some of the braces that should appear here.

We have noted by inspection that there were three successful trials. We can also do it in a different way, which appears to be more convenient when the number of trials is large.

```
>   add(outcomes[i],i=1..nops(outcomes));
                        3
```

We can perform more trials and find the fraction of successes.

```
>   stats[random,trial](1000):
    outcomes := map(F,[%]):
    evalf(add(outcomes[i],i=1..nops(outcomes))/1000);
                    .1960000000
```

1.6 MAPLE Session

MAPLE provides a number of procedures devoted to the calculus of probability and we have already used some of them. Except for the rand procedure which is available at any time, there are two packages containing such procedures. We first consider some useful procedures from the combinat package.

```
>   with(combinat);
```

[Chi, *bell*, binomial, *cartprod*, *character*, *choose*, *composition*, *conjpart*, *decodepart*, *encodepart*, *fibonacci*, *firstpart*, *graycode*, *inttovec*, *lastpart*, *multinomial*, *nextpart*, *numbcomb*, *numbcomp*, *numbpart*, *numbperm*, *partition*, *permute*, *powerset*, *prevpart*, *randcomb*, *randpart*, *randperm*, *stirling1*, *stirling2*, *subsets*, *vectoint*]

M 1.1 *Six balls are drawn without replacement from a box containing 5 white, 3 red and 2 green balls. Show some possible outcomes.*

```
>   for i from 1 to 7 do
    randcomb([w,w,w,w,w,r,r,r,g,g], 6);
    od;
```

$$[w, w, w, w, r, g]$$
$$[w, w, w, w, r, r]$$
$$[w, w, r, r, g, g]$$
$$[w, w, w, w, g, g]$$
$$[w, w, w, w, r, r]$$
$$[w, w, r, r, g, g]$$
$$[w, r, r, r, g, g]$$

M 1.2 *Continuation of Exercise M 1.1. What is the probability that we take exactly 3 white balls out? What is the probability that we will have 3 white, 2 red (and hence 1 green) balls?*

The answer to the first question:
```
>  binomial(5,3)*binomial(5,3)/binomial(10,6);
```
$$\frac{10}{21}$$

The answer to the second question:
```
>  binomial(5,3)*binomial(3,2)* binomial(2,1)/binomial(10,6);
```
$$\frac{2}{7}$$

M 1.3 *Continuation of Exercises M 1.1 and M 1.2. Answer the problem assuming that the balls are taken out with replacement.*

The answer to the first question:
```
>  binomial(6,3)*5^3*5^3/10^6;
```
$$\frac{5}{16}$$

The answer to the second question:
```
>  binomial(6,3)*binomial(3,2)*binomial(1,1)*5^3*3^2*2^1/
   10^6;
```
$$\frac{27}{200}$$

The same can be achieved in a simpler way:
```
>  multinomial(6,3,2,1)*5^3*3^2*2^1/10^6;
```
$$\frac{27}{200}$$

M 1.4 *Find the probabilities of all possible outcomes in the preceding problem.*

```
>  for i from 0 to 6 do
   for j from 0 to 6 - i do
   k := 6 - i - j:
   p[i,j] := multinomial(6,i,j,k)*5^i*3^j*2^k/10^6
   od
   od:
```

For example, the probability that all balls are green is:
```
>  p[0,0];
```
$$\frac{1}{15625}$$

And the probability that all balls are white is remarkably larger:
```
> p[6,0];
```
$$\frac{1}{64}$$

M 1.5 *Write down all possible cases when one is tossing four coins.*

```
> coin := [H,T]:
  alist := NULL:
  CT := cartprod([seq(coin,i=1..4)]):
  while not CT[finished] do
  alist := alist,CT[nextvalue]() od:
  alist;
```

$[H, H, H, H], [H, H, H, T], [H, H, T, H], [H, H, T, T], [H, T, H, H],$
$[H, T, H, T], [H, T, T, H], [H, T, T, T], [T, H, H, H], [T, H, H, T],$
$[T, H, T, H], [T, H, T, T], [T, T, H, H], [T, T, H, T], [T, T, T, H],$
$[T, T, T, T]$

For example, three repetitions might be as follows:
```
> randcomb([alist],3);
```
$$[[T, T, H, T], [T, T, T, H], [T, T, T, T]]$$

1.7 Exercises

E 1.1 Construct a probability space modeling the tossing of two dice if one knows the maximum number of pips obtained.

E 1.2 Construct a probability space modeling the tossing of two dice, if one knows the difference of the pips. Distinguish the cases: (a) the order of dice is considered, (b) the order of dice is not considered.

E 1.3 Prove that $P(\emptyset) = 0$. **Hint.** $\Omega = \Omega \cup \sum_{i=1}^{\infty} \emptyset$.

E 1.4 Prove Properties 1 through 5 in Proposition 1.1.1
  Hint. A finite union of sets may be considered as an infinite union by adjoining the empty set infinitely many times.
  Hint. $B = A \cup (B \setminus A)$, $A \cup B = (A \setminus B) \cup (B \setminus A) \cup (A \cap B)$.

E 1.5 Prove that $P(\bigcap_{n=0}^{\infty} A_n) = 1$, provided $P(A_n) = 1$ for any n.

E 1.6 Prove that for $A, K \in \Sigma$ with $P(K) = 1$ we have $P(A) = P(A \cap K)$.
  Hint. $A = (A \cap K) \cup (A \setminus K)$.

E 1.7 A population of fish with unknown magnitude N in a lake contains 20 marked fish. In a fish net of 30 fish 5 marked fish have been found. Find the size N of the population for which the above data are most likely to occur.

Comment. The quantity N serves as an estimator of the actual size of the population. The method used is called the maximum likelihood method.

E 1.8 What is the probability that, when we randomly put 5 letters in 5 addressed envelopes, we will have: (a) every letter in the appropriate envelope, (b) exactly r letters in the appropriate envelopes, $r = 0, 1, \ldots, 5$, (c) at least one letter in the appropriate envelope?

E 1.9 Prove that a probability space having probabilities of all the elementary events mutually equal and positive is finite.

E 1.10 A said that B said that C told a lie. What is the probability that C actually told a lie if one knows that each of the three tells lies two thirds of the time?

E 1.11 A couple has reached the following compromise. If on a given day the husband washes the dishes, then on the following day a toss of a coin decides who washes. If on a given day the wife washes the dishes, then the following day the husband does. A coin is used to decide who washes on the first day. Find: (a) The probability that the husband will wash the dishes on the third day. (b) The probability that the husband washed the dishes during first three days if we know he washed on the fourth day. (c) The probability that during the first week the couple will wash the dishes alternately.

E 1.12 We roll 10 dice again and again. Compute the probability that during n trials at least once we will get "6" on the all 10 dice. Using MAPLE get the numerical values for n = 1 000, 10 000, and 100 000.

E 1.13 Borel-Cantelli Lemma. For a sequence of events A_i, $i = 1, 2, 3, \ldots$, in a probability space (Ω, Σ, P) define the event A as:

$$A = \bigcap_{n=1}^{\infty} \bigcup_{i=n}^{\infty} A_i.$$

Prove:

1. If the series $\sum_{i=1}^{\infty} P(A_i)$ is finite, then $P(A) = 0$.
2. If the series $\sum_{i=1}^{\infty} P(A_i)$ is not finite and the events A_i are independent, then $P(A) = 1$.

Hint. In the proof of the second part use property from Exercise E 1.5 and the inequality $1 - x \leq e^{-x}$, for all $x \in \mathbb{R}$.

E 1.14 Using Borel-Cantelli Lemma prove that if one rolls 10 dice infinitely many times, then one will get "6" on the all 10 dice in infinite many trials.

Hint. Let A_i denote that in the i-th trial we get "6" on the all 10 dice. Then:

$$\omega \in \bigcap_{n=1}^{\infty} \bigcup_{i=n}^{\infty} A_i \iff \omega \text{ belongs to infinite many } A_i's$$

2 Measure and Integral

This Chapter provides a short introduction to measure theory and the theory of the Lebesgue integral. Avoiding as much as possible technical details, we present the basic ideas and also list the basic properties of measure and integral that will be used in this book. Let us note here that probability $P(A)$ defined in Chapter 1 is nothing else but the measure of event A. Another basic probability concept, the mathematical expectation, established later in the book is nothing else but the integral with respect to the probability measure. To some extent, calculus of probability, theory of stochastic processes (including stochastic differential equations) and mathematical statistics can be thought of as parts of the theory of measure and integration.

In spite of our efforts to simplify as much as possible, this chapter is perhaps the most demanding of the whole book. Its contents form the basis of much of what follows. Since many of the ideas may seen more natural, or even obvious, once the reader has seem them being used, we urge readers to at least peruse the chapter now and then return to it as they work their way through the rest of the book.

2.1 Measure

The reader has probably had a lot of practice in computing the length of segments, the length of arcs of a circle or even the length of more complicated curves. Also, he has surely computed the area of various plane figures and the area of some surfaces in the space as well as the volume bounded by such surfaces. It is likely, however, that the reader did not think much about the nature of these concepts. Does it make any sense? The meaning of length, area and volume seems to be so obvious. Let us think about this now and try to answer two simple questions.

Does every subset of the straight line have a length?

Does every figure on the plane have an area?

Well, this may seem trivial. We think the answer is yes, yet we are wrong. There exist plane figures for which one can not determine the area in a reasonable way and there exist subsets of the line with no meaningful length. In order to gain some initial understanding, we shall first analyse the following problem. Consider the set of rationals contained in the unit interval $[0,1]$ and the set of irrationals from this interval. What are the lengths of both these sets? Note that the sets here are disjoint and their union has length one. These hints are not helpful for us at the moment, so we feel we are not able to give any reasonable answer unless we clearly define what the length of the set is.

First, we will discuss some theoretical points and then return to our questions.

Try to imagine what common properties length, area and volume have, when these notions are understood in the usual every day meaning. First of all, each of them is a function which assigns real numbers (or infinity) to sets. The function takes nonnegative values and is additive, i.e. its value for the disjoint union $A \cup B$ is the same as the sum of values taken by the function for A and for B. Besides, the length of the interval $[0, 1]$, the area of the square $[0, 1]^2$ and the volume of the cube $[0, 1]^3$ are each equal to 1. Also, these functions are invariant under translations. Now, we are asking whether the above functions are defined for all subsets of the line, of the plane or of the space, respectively. We feel that they are defined for at least a broad class of such subsets. For instance, the area is defined for all triangles, rectangles, disks, circles (the last takes value zero), and much more. The above remarks make it easier to understand the following formal definition.

2.1.1 Formal Definitions

Assume we are given a nonempty set Ω. Denote by $\mathcal{P}(\Omega)$ the set of all subsets of Ω and consider a family Σ of subsets of Ω, so $\Sigma \subset \mathcal{P}(\Omega)$.

Definition 2.1.1 *We say that Σ is a σ-algebra if*

1. $\Omega \in \Sigma$,
2. *If $A_1, A_2, A_3, \ldots \in \Sigma$, then $\bigcup_{i=1}^{\infty} A_i \in \Sigma$.*
3. *If $A, B \in \Sigma$, then $A \setminus B \in \Sigma$.*

Note that we consider countably infinite unions of sets here. The above definition implies that the empty set belongs to a σ-algebra, $\emptyset \in \Sigma$, and that Σ is closed with respect to countably infinite intersections, i.e.

$$\text{If } A_1, A_2, A_3, \ldots \in \Sigma, \text{ then } \bigcap_{i=1}^{\infty} A_i \in \Sigma.$$

Instead of saying that a set A belongs to σ-algebra Σ we often say that A is Σ-measurable, or just measurable, if it is clear which σ-algebra is under discussion. The pair (Ω, Σ) is called a *measurable space*.

Trivial examples of σ-algebras are the family $\mathcal{P}(\Omega)$ of all subsets of Ω itself or the family $\{\emptyset, \Omega\}$. On the other hand, there are many classes of sets that are not σ-algebras. For example, the family of all open sets in $\mathbb{R}^n$ is not a σ-algebra as the last condition of the above definition is not satisfied.

Example 2.1.1 Let $S_1, \ldots S_r$ be a partition of a set Ω, i.e. $\bigcup_{i=1}^{r} S_i = \Omega$ and the S_i are pairwise disjoint. We define Σ to be a family of all unions of sets S_i. It is clear that Σ is a σ-algebra. Note that we have already encountered the above situation in Chapter 1 when we considered Ω as the set of all pairs (i, j), $i, j = 1, \ldots, 6$ and $S_k = \{(i, j) \in \Omega : i + j = k\}$, $k = 2, \ldots, 12$ (see page 2).

Example 2.1.2 An important σ-algebra is a family of so called Borel sets denoted by $\mathcal{B}(\mathbb{R}^n)$ and defined as the smallest σ-algebra containing all of the open sets of the space $\mathbb{R}^n$. Thus any open set is a Borel set, while any closed set is a Borel set since it is the complement of an open set. Moreover, singleton sets are Borel sets and any finite or countable sets are too. Hence, the set $\mathbb{Q}$ of rationals and its complement $\mathbb{R} \setminus \mathbb{Q}$, the set of irrationals, are Borel sets in $\mathbb{R}$.

Example 2.1.3 Lebesgue measurable sets form a slightly larger class $\mathcal{L}(\mathbb{R}^n)$ than the Borel sets $\mathcal{B}(\mathbb{R}^n)$. We do not state here a formal definition, but note that the class $\mathcal{L}(\mathbb{R}^n)$ is built up from the class $\mathcal{B}(\mathbb{R}^n)$ by adding to the latter class all subsets contained in very thin (in a certain definite sense) sets from $\mathcal{B}(\mathbb{R}^n)$. In particular, there exist Lebesgue measurable sets that are not Borel sets.

Now we are ready to state the definition of measure. Let (Ω, Σ) be a measurable space and let $\mu : \Sigma \longrightarrow \mathbb{R} \cup \{\infty\}$ be a function.

Definition 2.1.2 *We say that the function μ is a measure if*

1. *For all sets $A \in \Sigma$, $\mu(A) \geq 0$.*
2. *$\mu(\emptyset) = 0$.*
3. *If sets $A_1, A_2, A_3, \ldots \in \Sigma$ are pairwise disjoint, then*

$$\mu\left(\bigcup_{i=1}^{\infty} A_i\right) = \sum_{i=1}^{\infty} \mu(A_i).$$

There are many interesting examples of measures arising in probability theory. We saw some of them in Chapter 1 and we will discuss others later in this book. Here we just note that length, area and volume are measures on the straight line, the plane and in space, respectively. Namely, we state without giving the long proof the following:

Theorem 2.1.1 *Let the integer n be the dimension of the space $\mathbb{R}^n$. There exists a unique measure*

$$\mu = \mu_n : \mathcal{L}(\mathbb{R}^n) \longrightarrow \mathbb{R} \cup \{\infty\}$$

satisfying the following conditions:

1. *$\mu([0, 1]^n) = 1$, where $[0, 1]^n$ denotes the n-th Cartesian product of $[0, 1]$.*

2. *For every $A \in \mathcal{L}(\mathbb{R}^n)$ and every $p \in \mathbb{R}^n$*

$$\mu(A + p) = \mu(A)$$

(which means invariance of the measure μ with respect to translations).

The measure μ in the above Theorem is called the Lebesgue measure. Note that for $n = 1$, 2, 3 the Lebesgue measure gives, respectively, length, area and volume.

The Lebesgue measure has several useful properties. We list two of them.

Proposition 2.1.1 *The n-dimensional Lebesgue measure has the following properties:*

1. *For any Borel sets $A \subset \mathbb{R}^n$, $B \subset \mathbb{R}^m$ we have*

$$\mu_{n+m}(A \times B) = \mu_n(A)\mu_m(B).$$

2. *For any linear map $f : \mathbb{R}^n \longrightarrow \mathbb{R}^n$ of the form $f(x) = Ax$, where A is a square $n \times n$ matrix we have:*

$$\mu_n(f(K)) = |\det A|\, \mu_n(K),$$

where $f(K) = \{f(x) : x \in K\}$ is the image of the Lebesgue measurable set K.

The Lebesgue measure defined in the space $\mathbb{R}^n$ can be transported onto some other sets, for example a curve or a surface. We are not going to develop the whole theory but we confine ourselves to an important and simple case where the set is the unit circle centered at the origin of the plane, usually denoted by $\mathbb{S}^1$. Consider the parametrisation of $\mathbb{S}^1$:

$$\varphi : [0, 2\pi) \ni t \longrightarrow \varphi(t) = (\cos t, \sin t) \in \mathbb{S}^1.$$

Now we can define the Lebesgue measurable subsets of $\mathbb{S}^1$ as images of the ones contained in the interval $[0, 2\pi)$ and for any such set $B \subset \mathbb{S}^1$, $B = \varphi(A)$ we define its measure m by $m(B) = \mu(A)$, where $\mu = \mu_1$ is the Lebesgue measure. Let us note that the measure m is nothing else but the length of the arc. In particular $m(\mathbb{S}^1) = \mu([0, 2\pi)) = 2\pi$.

2.1.2 Geometric Probability Revisited

Definition 2.1.3 *Let $\Omega \subset \mathbb{R}^n$ be a Borel set ($\Omega \in \mathcal{B}(\mathbb{R}^n)$) having its Lebesgue measure $\mu(\Omega)$ positive and finite. Define a σ-algebra Σ as follows:*

$$\Sigma = \{A \cap \Omega : A \in \mathcal{B}(\mathbb{R}^n)\}$$

and the probability

$$P(A \cap \Omega) = \frac{\mu(A \cap \Omega)}{\mu(\Omega)}.$$

It is easy to verify that the triple (Ω, Σ, P) is a probability space. $P(A)$ is said to be a geometric probability of the set A.

The geometric probability is applicable in cases when the elementary events can be interpreted as points in the space $\mathbb{R}^n$ and, roughly speaking, are equally likely to occur.

One can straightforwardly extend the definition of geometric probability to these cases when we have a natural measure that has a geometric meaning, like the length of the arc on the unit circle $\mathbb{S}^1$.

2.1.3 Properties of Measure

Conditions used in the definition of a measure imply many important properties of the measure. We list some of them below.

Proposition 2.1.2 *Let μ be a measure defined on a measurable space (Ω, Σ). Then*

1. *If the sets $A_1, A_2, A_3, \ldots A_N \in \Sigma$ are pairwise disjoint, then*

$$\mu\left(\bigcup_{i=1}^{N} A_i\right) = \sum_{i=1}^{N} \mu(A_i).$$

2. *If $A, B \in \Sigma$ and $A \subset B$, then*

$$\mu(B) = \mu(A) + \mu(B \setminus A).$$

3. *For any $A_1, A_2, A_3, \ldots \in \Sigma$,*

$$\mu\left(\bigcup_{i=1}^{\infty} A_i\right) \leq \sum_{i=1}^{\infty} \mu(A_i).$$

4. *For any $A_1 \subset A_2 \subset A_3 \subset \ldots$ such that $A_n \in \Sigma$,*

$$\lim_{n \to \infty} \mu(A_n) = \mu\left(\bigcup_{n=1}^{\infty} A_n\right).$$

5. *For any $A_1 \supset A_2 \supset A_3 \supset, \ldots$ such that $A_n \in \Sigma$ and $\mu(A_1) < \infty$,*

$$\lim_{n \to \infty} \mu(A_n) = \mu(\bigcap_{n=1}^{\infty} A_n).$$

. . .

Now, we can go back to our earlier questions. Does any subset of the straight line have its length determined? Does any figure on the plane have its area determined? The answer is NO! by the following:

Theorem 2.1.2 *There exist subsets of $\mathbb{R}^n$ that are not Lebesgue measurable, in other words $\mathcal{L}(\mathbb{R}^n) \subsetneq \mathcal{P}(\mathbb{R}^n)$.*

There is no measure defined on the σ-algebra $\mathcal{P}(\mathbb{R}^n)$ which is equal to the Lebesgue measure when restricted to $\mathcal{L}(\mathbb{R}^n)$.

We omit the sophisticated proof and admit that it is not easy to visualise sets that are not Lebesgue measurable. The best known sets of this kind are the so called Vitali sets. It is worthwhile to mention here that $\mathcal{L}(\mathbb{R}^n)$ contains all sets that are subsets of Borel sets that have zero Lebesgue measure.

We also asked about the length of the set of all rationals in the unit interval $[0, 1]$. This set is a countable union of singletons. One can easily show that the Lebesgue measure of a singleton set is zero, so our set also has measure zero. Hence, its complement, the set of all irrationals in $[0, 1]$ has measure 1.

We make two final remarks.

From a naive point of view it may seen strange to consider infinite unions and intersections of sets in the definitions of σ-algebra and measure, while in "real life" we deal mostly with finite situations. This assumption however makes it possible to consider broad classes of sets as measurable, including the two considered above.

In addition, Theorem 2.1.1 implies the existence of a measure of the σ-algebra of Borel sets. It is the restriction of the Lebesgue measure μ.

2.2 Integral

2.2.1 The Riemann Integral

One of the most important mathematical concepts is integral. The reader will recall the definite or Riemann integral of elementary calculus,

$$\int_a^b f(x)\, dx,$$

where f is a scalar function defined on the interval $[a, b]$ and, recalling in particular, that the integral is just the area of the figure bounded by the graph of f, the X axis, and the lines $x = a$ and $x = b$, if f takes nonnegative values and $a < b$. If the function f is negative, then $\int_a^b f(x)\, dx$ is negative and its value is equal to minus the area of this figure.

Omitting some details, we recall that the definite integral is the limit of a sequence of sums

$$\int_a^b f(x)\,dx = \lim \sum_{i=1}^n f(\xi_i)(x_i - x_{i-1}),$$

where x_i are the points of the partition $a = x_0 < x_1 < \ldots < x_n = b$ and the $\xi_i \in [x_{i-1}, x_i]$ are arbitrarily chosen points. The limit is understood in such a way that n goes to infinity, while the lengths of all intervals (x_{i-1}, x_i) tend uniformly to zero. One requires the limit to be the same for all sequences of such partitions and any choice of evaluation points ξ_i. A geometric meaning of this definition is shown in the figure on page 40.

MAPLE ASSISTANCE 15

Compute a few approximation sums and compare them with the exact value of the definite integral $\int_0^\pi (1 + x \sin x)\,dx$.

Since we shall use procedures from the student package, we should read it in.

```
>  with(student);
```

[D, Diff, Doubleint, Int, Limit, Lineint, Product, Sum, Tripleint,
  changevar, combine, completesquare, distance, equate, extrema, integrand,
  intercept, intparts, isolate, leftbox, leftsum, makeproc, maximize,
  middlebox, middlesum, midpoint, minimize, powsubs, rightbox,
  rightsum, showtangent, simpson, slope, summand, trapezoid, value]

Define the following function:

```
>   f := x -> 1+x*sin(x);
```
$$f := x \rightarrow 1 + x \sin(x)$$

Calculate Riemann sums using the partition determined by the points $x_i = i\dfrac{\pi}{12}$, $i = 0, \ldots, 12$ and assume intermediate points ξ_i being, respectively, the left end points, the middle points and the right end points of the intervals $[x_{i-1}, x_i]$.

```
>  leftsum(f(x), x=0..Pi,12);
```
$$\frac{1}{12}\,\pi\,\left(\sum_{i=0}^{11}\left(1 + \frac{1}{12}\,i\,\pi\,\sin(\frac{1}{12}\,i\,\pi))\right)\right)$$

```
>  evalf(%);
```
$$6.265221330$$

```
>  middlesum(f(x), x=0..Pi,12);
```
$$\frac{1}{12}\,\pi\,\left(\sum_{i=0}^{11}\left(1 + \frac{1}{12}\,(i + \frac{1}{2})\,\pi\,\sin(\frac{1}{12}\,(i + \frac{1}{2})\,\pi))\right)\right)$$

```
> evalf(%);
```
$$6.292174999$$
```
> rightsum(f(x), x=0..Pi,12);
```
$$\frac{1}{12}\,\pi\,\left(\sum_{i=1}^{12}(1+\frac{1}{12}\,i\,\pi\sin(\frac{1}{12}\,i\,\pi))\right)$$
```
> evalf(%);
```
$$6.265221330$$

We can visualise these results, for example:
```
> middlebox(f(x), x=0..Pi,12,xtickmarks=3);
```

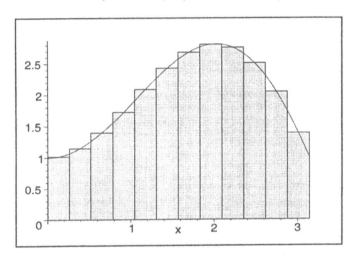

Now check the "true" value of the integral.
```
> int(f(x),x=0..Pi);
```
$$2\,\pi$$

We can see that the approximation sum based on the middle points is more accurate. So, compute a similar sum using a partition with 36 points.
```
> evalf(middlesum(f(x), x=0..Pi,36));
```
$$6.284182389$$

· · ·

Intuition tells us that for most known functions the above definition determines the definite integral. In fact, one can prove that the integral $\int_a^b f(x)\,dx$ exists and is finite for a continuous function f. The same applies for bounded functions having at most a countable number points of discontinuity. There are, however, functions for which the definite integral cannot be defined.

Example 2.2.1 Consider the function $f : [0,1] \longrightarrow \mathbb{R}$ defined as

$$f(x) = \begin{cases} 0 \text{ for } x \in \mathbb{Q} \cap [0,1] \\ 1 \text{ for } x \in [0,1] \setminus \mathbb{Q} \end{cases}$$

Fix a partition $0 = x_0 < x_1 < \ldots < x_n = 1$. Taking points ξ_i to be rationals we get

$$\sum_{i=1}^{n} f(\xi_i)(x_i - x_{i-1}) = 0.$$

For irrational points ξ_i we get

$$\sum_{i=1}^{n} f(\xi_i)(x_i - x_{i-1}) = 1.$$

As n approaches infinity we get a limit of 0 or 1, which means the integral $\int_0^1 f(x)\,dx$ does not exist.

2.2.2 The Stieltjes Integral

A possible generalization of the Riemann integral is the Stieltjes integral, which is sometimes called Riemann-Stieltjes integral and is defined as follows. Let f and g be functions defined on the same bounded interval $[a,b]$. We define

$$\int_a^b f(x)\,dg(x) = \lim \sum_{i=1}^{n} f(\xi_i)(g(x_i) - g(x_{i-1})),$$

where x_i are the points of the partition $a = x_0 < x_1 < \ldots < x_n = b$ and the $\xi_i \in [x_{i-1}, x_i]$ are arbitrarily chosen points. The limit is understood in such a way that n goes to infinity, while the lengths of all intervals (x_{i-1}, x_i) tend uniformly to zero. One requires the limit to be the same for all sequences of such partitions and any choice of evaluation points ξ_i.

Clearly, the Stieltjes integral is the same as the Riemann integral for the identity function g, $g(x) = x$. The Stieltjes integral exists for a broad class of functions. In particular, if f is a continuous function and g is monotone, then the Stieltjes integral exists. If, additionally, g is differentiable, then

$$\int_a^b f(x)\,dg(x) = \int_a^b f(x)g'(x)\,dx.$$

We will see, however, in Chapter 8 that the straightforward application of the Stieltjes integral for stochastic processes is not possible and then we will establish the concept of an Ito integral.

More general than the Riemann and Stieltjes integral is the so called Lebesgue integral or integral with respect to a measure. The main idea is to generalise the concept of definite integral to a broad class of real valued functions defined on any set with a distinguished σ-algebra and a measure rather than just on the real line. The concept of the Lebesgue integral is exceptionally useful to establish and examine mathematical expectation (see Chapter 4). In order to give a formal definition we need to introduce the concept of a measurable function.

2.2.3 Measurable Functions

Let Ω be a non-empty set, Σ be a σ-algebra on Ω and $f : \Omega \longrightarrow \overline{\mathbb{R}}$, where $\overline{\mathbb{R}} = \mathbb{R} \cup \{-\infty, \infty\}$.

Definition 2.2.1 *We say the function f is Σ-measurable (or just measurable), if for any Borel set, $B \in \mathcal{B}(\mathbb{R})$, its preimage is Σ-measurable, that is,*

$$f^{-1}(B) = \{\omega \in \Omega : f(\omega) \in B\} \in \Sigma.$$

One can show that we do not need to take all of the Borel sets in the above definition. Namely we have the following.

Theorem 2.2.1 *The following conditions are equivalent.*

1. *f is measurable.*
2. *For any open set $G \subset \mathbb{R}$, $f^{-1}(G) \in \Sigma$.*
3. *For any closed set $F \subset \mathbb{R}$, $f^{-1}(F) \in \Sigma$.*
4. *For any $y \in \mathbb{R}$, $\{\omega \in \Omega : f(\omega) < y\} \in \Sigma$.*
5. *For any $y \in \mathbb{R}$, $\{\omega \in \Omega : f(\omega) \leq y\} \in \Sigma$.*
6. *For any $y \in \mathbb{R}$, $\{\omega \in \Omega : f(\omega) > y\} \in \Sigma$.*
7. *For any $y \in \mathbb{R}$, $\{\omega \in \Omega : f(\omega) \geq y\} \in \Sigma$.*

We see that the measurability of a given function depends on the σ-algebra. When the σ-algebra contains just a few sets, there are not so many measurable functions, while if the σ-algebra is large, many more functions are measurable. Consider the two extreme cases: for $\Sigma = \{\emptyset, \Omega\}$ the only measurable functions are the constant functions, while for $\Sigma = \mathcal{P}(\Omega)$ all functions are measurable.

Important examples of measurable functions are indicators (more precisely, indicator functions) of measurable sets $A \in \Sigma$. They are functions defined as

$$I_A(x) = \begin{cases} 1, & \text{for } x \in A \\ 0, & \text{for } x \notin A. \end{cases}$$

In fact, the only possible preimages here are $\emptyset, A, \Omega \setminus A, \Omega$, which are all measurable.

Indicator functions are used to build more sophisticated measurable functions through basic numerical operations, taking the minimum or maximum of functions (or generally, infimum or supremum), limits and compositions with continuous functions. Namely, one can prove the following.

Theorem 2.2.2 *Measurable functions share the following properties.*

1. *The indicator function of a measurable set is measurable.*
2. *The sum, difference, product, quotient (if defined) of measurable functions is measurable.*
3. *The minimum and maximum of two measurable functions is measurable.*
4. *The supremum and infimum of a sequence of measurable functions is measurable.*
5. *The limit of a sequence of measurable functions is measurable.*
6. *The composition $g \circ f$ of a measurable function f with a continuous function g is measurable.*
   In particular the following functions are measurable:
7. *Step functions defined as*

$$f(x) = c_i, \text{ for } x \in A_i,$$

where $A_1, \ldots, A_k$ are measurable and pairwise disjoint sets with their union equal to the whole space Ω, and arbitrary numbers $c_1, \ldots, c_k$. They are just the sum of products of measurable functions with numbers

$$f(x) = \sum_{i=1}^{k} c_i I_{A_i}.$$

8. *The functions f^+ and f^- defined as*

$$f^+(x) = \max(f(x), 0), \qquad f^-(x) = -\min(f(x), 0),$$

if f is measurable.

When $\Omega = \mathbb{R}^n$, and we consider the σ-algebra to be $\mathcal{B}(\mathbb{R}^n)$ or $\mathcal{L}(\mathbb{R}^n)$, the class of measurable function is very large. In particular any continuous function is measurable since preimages of any open set by a continuous function is also open, and hence a Borel set.

We say that functions $f, g : \Omega \longrightarrow \overline{\mathbb{R}}$ are *equal almost everywhere*, writing this as $f = g$ a.e., if $\mu(\{\omega \in \Omega : f(\omega) \neq g(\omega)\}) = 0$. Similarly, $f \leq g$ a.e. means that $\mu(\{\omega \in \Omega : f(\omega) > g(\omega)\}) = 0$. A common practice is to omit the phrase almost everywhere while talking about equality or inequality of measurable functions. In the case where measure μ is a probability measure we will often use the phrase *with probability one*, w.p.1, instead of almost everywhere.

2.2.4 The Lebesgue Integral

We present the definition of the Lebesgue integral through a three step procedure.

Definition 2.2.2 *Let (Ω, Σ) be a measurable space, let $\mu : \Sigma \longrightarrow \mathbb{R} \cup \{\infty\}$ be a measure, and let $f : \Omega \longrightarrow \mathbb{R}$ be a function.*

1. *Assume f to be a nonnegative measurable step function. That is*

$$f(x) = c_i, \ for \ x \in A_i,$$

and define

$$\int_\Omega f \, d\mu = \sum_{i=1}^k c_i \mu(A_i).$$

2. *Assume f to be a nonnegative measurable function. One can show the existence of nonnegative measurable step functions f_n, $n = 1, 2, 3, \ldots$ such that*

$$f_n(x) \le f_{n+1}(x),$$

for all n and x, and

$$\lim_{n \to \infty} f_n(x) = f(x).$$

For such a sequence f_n define

$$\int_\Omega f \, d\mu = \lim_{n \to \infty} \int_\Omega f_n \, d\mu.$$

One can prove that the above limit exists (finite or infinite) and does not depend on the choice of the sequence f_n.

3. *Assume f to be any measurable function. Then*

$$f = f^+ - f^-,$$

where f^+, f^- are nonnegative measurable functions. Define

$$\int_\Omega f \, d\mu = \int_\Omega f^+ \, d\mu - \int_\Omega f^- \, d\mu$$

when the right hand side is determined (i.e. is not of the form $\infty - \infty$).

We comment on the above definition. It may happen that the integral is infinite or does not exist, even if the function f is finite. In Point 1, if the measure of at least one A_i is infinite and the corresponding constant c_i (the height of the bar of the base A_i) is positive then the integral is infinite (note, in measure theory one assumes that $0 \times \infty = 0$). The integral defined in Point 2 may be infinite, even though all $\int_\Omega f_n \, d\mu$ are finite. The integral defined in

Point 3 for an arbitrary measurable function may not exist, which happens if both integrals $\int_\Omega f^+ \, d\mu$ and $\int_\Omega f^- \, d\mu$ are infinite. In all other cases the integral exists (finite or infinite). Taking this into account we say that function f is *integrable* if it is measurable and the integral $\int_\Omega f \, d\mu$ exists. The function is *summable*, written $f \in L^1(\Omega)$, if it is measurable and the integral $\int_\Omega f \, d\mu$ is finite.

The crucial step in the definition of the Lebesgue integral is Point 2. It actually says that the integral is a limit of approximate sums, as in the definition of the Riemann integral. The basic difference is that the measure μ may be much more general than the length and even if we consider μ to be length, the sets A_i need not be intervals and their measure may be infinite.

Example 2.2.2 We will interpret Point 2 from the above definition of integral. For the function $f(x) = e^{-|x|}$ consider the integral $\int_\mathbb{R} f \, d\mu$, where μ is a 1-dimensional Lebesgue measure. We have to find a sequence of measurable step functions f_n, such that for every point $x \in \mathbb{R}$, $f_n(x) \le f_{n+1}(x)$ and $\lim_{n\to\infty} f_n(x) = f(x)$. Let n be of the form $n = 2^k$, $k = 1, 2, 3, \ldots$. For $i = 1, \ldots n - 1$ define points $a_i > 0$ such that $f(a_i) = \frac{i}{n}$ and put $a_0 = \infty$, $a_n = 0$. Consider the function

$$f_n(x) = \frac{i-1}{n}, \text{ for } x \in (-a_{i-1}, -a_i] \cup [a_i, a_{i-1}), \quad i = 1, \ldots, n,$$

where we mean $\frac{0}{n} = \infty$. Clearly the sequence f_{2^k} fulfills the requirements and thus we have

$$\int_\mathbb{R} f \, d\mu = \lim_{k\to\infty} f_{2^k} \, d\mu$$

$$= \lim_{k\to\infty} \sum_{i=1}^{2^k} f_{2^k}(a_i)\mu((-a_{i-1}, -a_i] \cup [a_i, a_{i-1}))$$

$$= 2 \lim_{k\to\infty} \sum_{i=2}^{2^k} \frac{i-1}{2^k}(a_i - a_{i-1})$$

We visualize the above situation.

MAPLE ASSISTANCE 16
```
>  f := x ->exp(-abs(x));
```
$$f := x \to e^{(-|x|)}$$

```
> plot(f, -3..3);
```

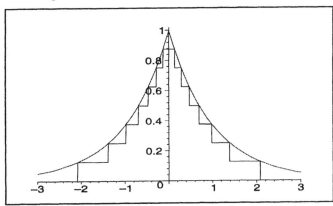

```
> n := 8:
> for i from 1 to n-1 do
  a[i] := fsolve(f(x) = i/n,x = 0..5)
  od:
  a[n] := 0:
> seq(evalf(a[i],4),i=1..8);
        2.079, 1.386, .9808, .6931, .4700, .2877, .1335, 0.
```

For simplicity we will write g instead of f_n.

```
> g := x -> piecewise( seq( op( [
  a[i] <= x and x < a[i-1],(i-1)/n,
  -a[i-1] < x and x <= - a[i],(i-1)/n] ),
  i=2..n) ):
> plot([f,g],-3..3,color = [RED,BLUE]);
```

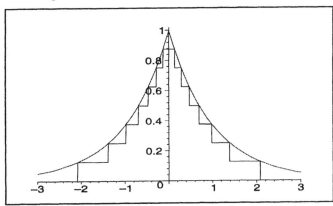

We can compute the integral $\int_{\mathbb{R}} g \, d\mu$.

```
> add(g(a[i])*2*(a[i-1] - a[i]), i = 2..n);
               1.507732358
```

See MAPLE Session later in the chapter for further discussion.

Example 2.2.3 Fix real numbers $a_i \geq 0$, $i = 1, 2, 3, \ldots$. We will show that the series $\sum_{i=1}^{\infty} a_i$ can be interpreted as an integral. Consider f, a real valued function defined on the set of natural numbers $\mathbb{N}$ as $f(i) = a_i$, for $i \in \mathbb{N}$. Consider the *counting measure* μ defined on the σ-algebra of all subsets of $\mathbb{N}$, i.e. $\mu(A) = \sharp A$ – the number of elements of A, possibly infinite. Define functions $f_n : \mathbb{N} \longrightarrow \mathbb{R}$ as $f_n(i) = f(i) = a_i$, for $i \leq n$ and $f_n(i) = 0$ otherwise. Clearly, f_n are step functions, $f_n(i) \leq f_n(i+1)$ and $f_n(i) \longrightarrow f(i)$ for all $i \in \mathbb{N}$. We then have

$$\int_{\mathbb{N}} f \, d\mu = \lim_{n \to \infty} \int_{\mathbb{N}} f_n \, d\mu = \lim_{n \to \infty} \sum_{i=1}^{n} f_n(i) \mu(\{i\}) = \lim_{n \to \infty} \sum_{i=1}^{n} a_i = \sum_{i=1}^{\infty} a_i$$

See also Exercise **E 2.9**.

Various notation is used for the Lebesgue integral. For example, if we want to indicate that the function f depends on a variable, say x, we often write

$$\int_{\Omega} f(x) \, d\mu(x) = \int_{\Omega} f(x) \, \mu(dx) = \int_{\Omega} f \, d\mu.$$

We can define the integral for a function $f : A \longrightarrow \overline{\mathbb{R}}$, where A is any measurable set as follows. The sets from σ-algebra Σ contained in A form a σ-algebra on A, say Σ_A. We can consider the restriction of the measure μ to Σ_A. It is a measure denoted by μ_A. One can define the integral with respect to this measure, so we define

$$\int_A f \, d\mu = \int_A f \, d\mu_A.$$

Alternatively, one can extend the function f to the whole space Ω as

$$\tilde{f}(x) = f(x) I_A(x) = \begin{cases} f(x), & \text{for } x \in A \\ 0, & \text{for } x \notin A \end{cases}$$

and define

$$\int_A f \, d\mu = \int_{\Omega} \tilde{f} \, d\mu.$$

2.3 Properties of the Integral

We state some basic properties of the Lebesgue integral needed in the sequel.

Theorem 2.3.1 *Let (Ω, Σ) be a measurable space and $\mu : \Sigma \longrightarrow \mathbb{R} \cup \{\infty\}$ be a measure.*

1. Let $f, g : \Omega \longrightarrow \overline{\mathbb{R}}$ be summable functions and let $\alpha, \beta \in \mathbb{R}$. Then, $\alpha f + \beta g$ is summable and

$$\int_\Omega (\alpha f + \beta g) \, d\mu = \alpha \int_\Omega f \, d\mu + \beta \int_\Omega g \, d\mu.$$

2. Let $f, g : \Omega \longrightarrow \overline{\mathbb{R}}$ be integrable functions. If $f \leq g$ a.e. then $\int_\Omega f \, d\mu \leq \int_\Omega g \, d\mu$.

3. Let $f, g : \Omega \longrightarrow \overline{\mathbb{R}}$ be integrable functions. If $f = g$ a.e. then $\int_\Omega f \, d\mu = \int_\Omega g \, d\mu$.

4. Let $f : \Omega \longrightarrow \overline{\mathbb{R}}$ be an integrable function. If $f \geq 0$ a.e. and $\int_\Omega f \, d\mu = 0$ then $f = 0$ a.e.

5. Let $f : \Omega \longrightarrow \overline{\mathbb{R}}$ be an integrable function. Then

$$\left| \int_\Omega f \, d\mu \right| \leq \int_\Omega |f| \, d\mu$$

6. Let $f, g : \Omega \longrightarrow \overline{\mathbb{R}}$ be integrable functions and $\int_\Omega f^2 \, d\mu < \infty$, $\int_\Omega g^2 \, d\mu < \infty$. Then

$$\int_\Omega (f \cdot g)^2 \, d\mu \leq \int_\Omega f^2 \, d\mu \cdot \int_\Omega g^2 \, d\mu \qquad (2.1)$$

and the equality holds if and only if $f = 0$ or $g = 0$ or there is an α such that $g = \alpha f$.

7. Let $f_n : \Omega \longrightarrow \overline{\mathbb{R}}$ be a sequence of integrable functions and let

$$f(\omega) = \lim_{n \to \infty} f_n(\omega)$$

which exists for all $\omega \in \Omega$. Assume alternatively that
either all $f_n \geq 0$ a.e. and $f_n(\omega) \leq f_{n+1}(\omega)$ for all n and ω,
or $|f_n| \leq g$ for all n, where g is a summable function.
Then

$$\lim_{n \to \infty} \int_\Omega f_n \, d\mu = \int_\Omega f \, d\mu.$$

8. Let $f : \Omega \longrightarrow \overline{\mathbb{R}}$ be an integrable function and let $A_1, A_2, A_3, \ldots \in \Sigma$ be pairwise disjoint sets. Then

$$\int_{\bigcup_{i=1}^\infty A_i} f \, d\mu = \sum_{i=1}^\infty \int_{A_i} f \, d\mu. \qquad (2.2)$$

2.3.1 Integrals with Respect to Lebesgue Measure

Consider the situation when $\Omega = \mathbb{R}^n$ and $\Sigma = \mathcal{B}(\mathbb{R}^n)$. A measurable function $f : \mathbb{R}^n \longrightarrow \overline{\mathbb{R}}$ in this case is said to be a Borel function. We often simplify notation in this case and write $\int_A f \, dx$ or $\int_A f(x) \, dx$ instead of $\int_A f \, d\mu$, for $A \in \mathcal{B}(\mathbb{R}^n)$.

The following theorems are often used.

Theorem 2.3.2 *Let $f : \mathbb{R}^n \longrightarrow \overline{\mathbb{R}}$ be a Borel function, nonnegative on a Borel set $A \subset \mathbb{R}^n$. Denote $A_f = \{(x,y) : x \in A, 0 \leq y \leq f(x)\}$. Then*

$$\mu_{n+1}(A_f) = \int_A f(x) \, d\mu_n(x).$$

Theorem 2.3.3 (Change of Variables) *Let $f : \mathbb{R}^n \longrightarrow \overline{\mathbb{R}}$ be a Borel function and let $\varphi : \mathbb{R}^n \longrightarrow \mathbb{R}^n$ be a diffeomorphism. Then, for any set $A \in \mathcal{B}(\mathbb{R}^n)$ we have*

$$\int_{\varphi^{-1}(A)} f(x) \, dx = \int_A f(\varphi^{-1}(x)) \, |Jac_x\varphi^{-1}| \, dx.$$

Here $Jac_x\varphi = \det\left[\frac{\partial\varphi_i}{\partial x_j}\right]$ denotes the Jacobian of a differentiable function f.

Theorem 2.3.4 (Fubini's Theorem) *Let $f : \mathbb{R}^n \times \mathbb{R}^m \longrightarrow \overline{\mathbb{R}}$ be a Borel function. Assume alternatively that either $f \geq 0$ a.e. or f is integrable.*
Then for any $A \in \mathcal{B}(\mathbb{R}^n)$ and $B \in \mathcal{B}(\mathbb{R}^m)$

$$\int_{A \times B} f(x,y) \, d\mu_{n+m}(x,y) = \int_A \left(\int_B f(x,y) \, d\mu_m(y) \right) d\mu_n(x). \qquad (2.3)$$

2.3.2 Riemann Integral Versus Lebesgue Integral

Consider the real line $\mathbb{R}$ with the σ-algebra $\mathcal{L}(\mathbb{R})$ of the Lebesgue measurable sets and fix a set A to be a bounded interval $[a,b]$. For a function $f : A \longrightarrow \mathbb{R}$ we have two integrals;

$$\text{The Riemann integral} - \int_a^b f(x) \, dx$$

and

$$\text{The Lebesgue integral} - \int_A f \, d\mu,$$

where μ is the Lebesgue measure. We ask if there are some connections between these two. Generally, there is no such connection, except for some interesting cases.

Example 2.3.1 In Example 2.2.1 we saw that the Riemann integral

$$\int_0^1 f(x)\, dx$$

did not exist when the function f was defined as

$$f(x) = I_{\mathbb{Q}\cap[0,1]}(x) = \begin{cases} 0 \text{ for } x \in \mathbb{Q}\cap[0,1] \\ 1 \text{ for } x \in [0,1]\setminus\mathbb{Q} \end{cases}.$$

However, the Lebesgue integral

$$\int_{[0,1]} f\, d\mu$$

does exist and equals 1, as the function f is the indicator of the measurable set $[0,1]\setminus\mathbb{Q}$ having Lebesgue measure 1.

On the other hand we have the following:

Theorem 2.3.5 *Let $a, b \in \mathbb{R}$, $a < b$. Let $f : [a,b] \longrightarrow \mathbb{R}$ be bounded and continuous in all points from $[a,b]\setminus K$, where K is countable or finite. Then both the Riemann integral and the Lebesgue integral exist and are equal. That is,*

$$\int_a^b f(x)\, dx = \int_{[a,b]} f\, d\mu.$$

2.3.3 Integrals Versus Derivatives

The following explains the fundamental relationship between Riemann integrals and derivatives.

Theorem 2.3.6 *Let $f : \Delta \longrightarrow \mathbb{R}$ be a continuous function on an interval $\Delta \subset \mathbb{R}$. Then*

1. *For any point $a \in \Delta$ the function F defined for $x \in \Delta$ by*

$$F(x) = \int_a^x f(s)\, ds$$

   is differentiable and its derivative $F'(x)$ at any point x is

$$F'(x) = f(x). \tag{2.4}$$

2. *For any function $F : \Delta \longrightarrow \mathbb{R}$ satisfying (2.4) we have*

$$\int_a^b f(x)\, dx = F(b) - F(a) \text{ for any } a, b \in \Delta. \tag{2.5}$$

2.4 Determining Integrals

Formula (2.5) allows us to compute Riemann integrals, and thus in many situations Lebesgue integrals. The problem reduces to finding a function F satisfying formula (2.4) (called an antiderivative). Such a function is not unique, although the difference $F(b) - F(a)$ is.

MAPLE ASSISTANCE 17

Compute an antiderivative F for the function $f(x) = x^2$ and then the Riemann integral $\int_{-1}^{3} f(x)\, dx$.

```
> f := unapply(x^2,x):
> F := unapply(int(f(x),x),x);
```

$$F := x \to \frac{1}{3}x^3$$

Note that MAPLE has printed out just one antiderivative. Any other antiderivative has the form $F(x) + c$, where c is a constant.

```
> F(3) - F(-1);
```

$$\frac{28}{3}$$

The MAPLE command "int" allows us to perform the above two steps in one.

```
> int(x^2,x=-1..3);
```

$$\frac{28}{3}$$

There exist functions for which the antiderivatives are hard to determine (although they exist in theory).

MAPLE ASSISTANCE 18

Compute $\int_{1}^{2} e^{x^2}\, dx$.

```
> int(exp(x^2),x = 1..2);
```

$$-\frac{1}{2}I\operatorname{erf}(2I)\sqrt{\pi} + \frac{1}{2}I\operatorname{erf}(I)\sqrt{\pi}$$

A non-elementary function erf and its values at the imaginary point I have appeared. MAPLE could not find the antiderivative in terms of elementary functions, i.e. functions that can be expressed as compositions of polynomials, exponential functions, trigonometric functions and their inverses. In such a situation MAPLE uses a non-elementary function (if it knows one) to evaluate the Riemann integral. Now, we can get an approximate value of our integral.

```
> evalf(%,15);
```

$$14.9899760196001$$

One can also compute some Lebesgue integrals of the functions of several variables. A common method is to reduce such a computation to a computation of two or more definite integrals. Fubini's Theorem (Theorem 2.3.4) is essential here.

Example 2.4.1 Compute

$$\int_A (x^2 + 2y)\, d(x, y)$$

denoted also as $\int_A (x^2 + 2y)\, dx dy$, where A is the rectangle $[1, 3] \times [2, 5]$ on the plane.

By Fubini's Theorem we have

$$\int_A (x^2 + 2y)\, d(x, y) = \int_{[1,3]} \left(\int_{[2,5]} (x^2 + 2y)\, dy \right) dx$$

$$= \int_1^3 \left(\int_2^5 (x^2 + 2y)\, dy \right) dx = \int_1^3 (x^2 y + y^2)\big|_2^5\, dx$$

$$= \int_1^3 (5x^2 + 25 - (2x^2 + 4))\, dx = \int_1^3 (3x^2 + 21)\, dx$$

$$= (x^3 + 21x)\big|_1^3 = 68.$$

The story is much more complicated if the set A we want to integrate over is not a rectangle. Some procedures, such as changing of variables, may be used to reduce the problem to integration over a rectangle. Still, if the set A is not regular enough, it may be impossible to find the integral.

In the following MAPLE Session we address some problems associated with the numerical computation of an integral.

2.5 MAPLE Session

M 2.1 *What formulas are used by the* student *package to calculate approximation sums of Riemann integrals?*

Delete any previous definition of f in the worksheet.

```
>  f := 'f':
```

We have the following:

```
>  leftsum(f(x), x = a..b,n);
```

$$\frac{(b - a) \left(\sum_{i=0}^{n-1} f(a + \frac{i(b - a)}{n}) \right)}{n}$$

```
> rightsum(f(x), x = a..b,n);
```

$$\frac{(b-a)\left(\sum_{i=1}^{n}\mathrm{f}(a+\frac{i(b-a)}{n})\right)}{n}$$

```
> middlesum(f(x), x = a..b,n);
```

$$\frac{(b-a)\left(\sum_{i=0}^{n-1}\mathrm{f}\left(a+\frac{(i+\frac{1}{2})(b-a)}{n}\right)\right)}{n}$$

M 2.2 *Compare graphically the above approximation sums of the Riemann integral.*

We have to fix some function f and the range of integration. For example, let $f(x) = x(4-x)$ for $0 \le x \le 4$.

```
> leftbox(x*(4 - x), x = 0..4, 6, xtickmarks=3);
```

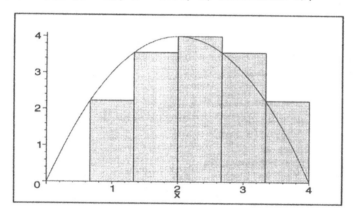

```
> rightbox(x*(4 - x), x = 0..4, 6, xtickmarks=3);
```

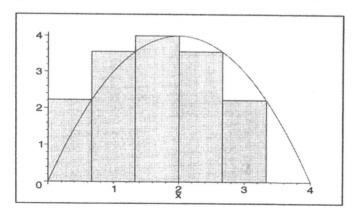

```
> middlebox(x*(4 - x), x = 0..4, 6, xtickmarks=3);
```

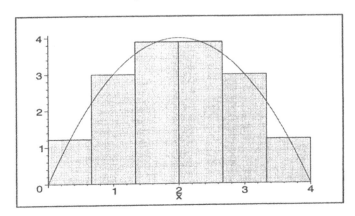

M 2.3 *Visualize the definition of the Lebesgue integral with respect to the Lebesgue measure over the whole real line for the function $f(x) = e^{-x^2}$.*

We take advantage of this example to show how to effectively built a procedure from the commands that have been already used (see page 46).

```
> f := x -> exp(-x^2):
```

The following procedure either will plot an approximating function along with function f or will compute the integral from the approximation.

```
> approx := proc(n,opt)
  local i, g, a:
  for i from 1 to n-1 do
  a[i] := fsolve(f(x) = i/n,x = 0..5)
  od:
  a[n] := 0: g := x -> piecewise( seq( op( [
  a[i] <= x and x < a[i-1],(i-1)/n,
  -a[i-1] < x and x <= -a[i],(i-1)/n] ),
  i=2..n) ):
  if opt = 0 then RETURN(plot([f,g],-3..3, color =
  [RED,BLUE])) fi:
  if opt = 1 then RETURN(add(g[n](a[i-1])*2*(a[i-1] - a[i]),
  i = 2..n) ) fi:
  end:
> plots[display]([approx(4,0), approx(8,0),
  approx(16,0), approx(32,0)],
  insequence=true);
```

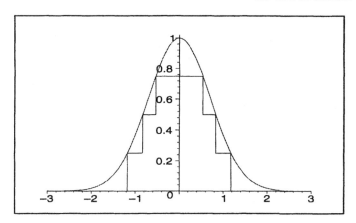

The above figure can be animated by clicking it with the mouse and using the context menu that appears.

We can also display all the plots.

```
>   plots[display](%,tickmarks=[0,0]);
```

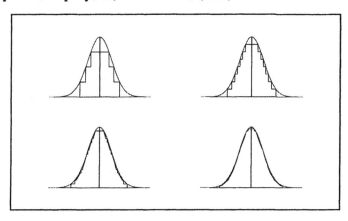

The corresponding integrals take values.

```
>   approx(4,1),approx(8,1),approx(16,1),approx(32,1);
        .6844573162, 1.146920120, 1.423951474, 1.581981494
```

What is the value of $\int_{\mathbb{R}} f \, d\mu$?

```
>   int(f(x),x = -infinity..infinity);
```
$$\sqrt{\pi}$$

```
>   evalf(%);
```
$$1.772453851$$

At last we compute its approximation with $n = 100$.
```
>  approx(100,1);
```
$$1.703772735$$

M 2.4 *Visualize the positive and negative parts f^+ and f^- for a given function f.*

Consider some function.
```
>  f := x -> (x^3 - 4*x)*sin(x-1/2):
>  fplus := x -> max(f(x),0):
>  fminus := x -> max(-f(x),0):
>  p := array(1..2,1..2):
>  p[1,1] := plot(f,-3..3):
>  p[1,2] := plot(fplus,-3..3):
>  p[2,1] := plot(fminus,-3..3):
>  p[2,2] := plot(fplus - fminus, -3..3):
>  plots[display](p,tickmarks=[2,1]);
```

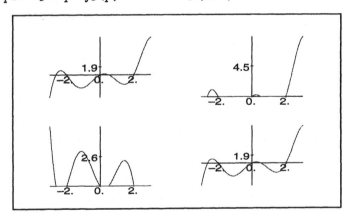

Note that we have used a different method of displaying plots than in the previous example.
```
>  evalf(int(fplus,-3..3));
```
$$5.441420955$$
```
>  evalf(int(fminus,-3..3));
```
$$6.443510285$$
```
>  evalf(int(f,-3..3));
```
$$-1.002089329$$

> `%%% - %%;`

$$-1.002089330$$

M 2.5 *Show the formulas used for approximating a definite integral by the trapezoidal method and the Simpson method; two classical methods for numerical integration, which were proposed as improvements of the approximation with the middle sums above.*

> `trapezoid(f(x), x = a..b,n);`

$$\frac{1}{2}\frac{(b-a)\left(f(a) + 2\left(\sum_{i=1}^{n-1} f(a + \frac{i\,(b-a)}{n})\right) + f(b)\right)}{n}$$

> `simpson(f(x), x = a..b,n);`

$$\frac{1}{3}\left((b-a)\left(f(a) + f(b) + 4\left(\sum_{i=1}^{1/2\,n} f(a + \frac{(2\,i - 1)\,(b-a)}{n})\right)\right.\right.$$
$$\left.\left. + 2\left(\sum_{i=1}^{1/2\,n-1} f(a + 2\,\frac{i\,(b-a)}{n})\right)\right)\right) / n$$

Actually, modern numerical integration methods are much more sophisticated (see below).

M 2.6 *Which method from the above is the best one to approximate the definite integral?*

We are not going to give a definite answer here but urge the reader to perform a series of experiments like the following.

Consider for example the following function f:
> `f := x -> x*exp(-x);`

$$f := x \to x\,e^{(-x)}$$

Fix a range of integration, plot the graph and find the integral.
> `therange := 0..4:`
> `plot(f,therange);`

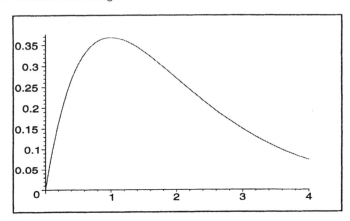

```
>  int(f(x),x=therange);
```
$$-5\,e^{(-4)} + 1$$

Its approximation is found below.
```
>  evalf(%);
```
$$.9084218056$$

We will compare the following methods.
```
>  method := [leftsum,rightsum,middlesum,trapezoid,simpson];
```
$$method := [leftsum,\ rightsum,\ middlesum,\ trapezoid,\ simpson]$$
```
>  seq(evalf(mm(f(x), x = therange,20)),mm = method);
```
$$.8975857572,\ .9122382682,\ .9101741916,\ .9049120128,\ .9083951859$$

Now, consider a more complicated function.
```
>  f := x -> sqrt(x) - cos(x)^2;
```
$$f := x \rightarrow \sqrt{x} - \cos(x)^2$$
```
>  therange := 0..3*Pi:
>  plot(f,therange);
```

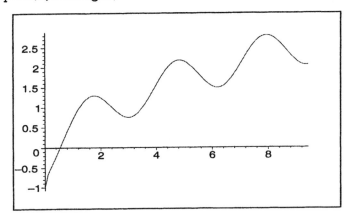

```
>  int(f(x),x=therange);
```

$$2\sqrt{3}\,\pi^{(3/2)} - \frac{3}{2}\,\pi$$

```
>  evalf(%);
```

$$14.57686504$$

```
>  seq(evalf(mm(f(x), x = therange,10)),mm = method);
```
$$12.95201568,\ 15.84540377,\ 14.62654956,\ 14.39870973,\ 14.50258953$$

Increase the number of subintervals.

```
>  seq(evalf(mm(f(x), x = therange,40)),mm = method);
```
$$14.19216880,\ 14.91551582,\ 14.58345218,\ 14.55384230,\ 14.56757986$$

Compare the above with the value obtained by a MAPLE default numerical method for integral computation.

```
>  evalf(Int(f(x), x = therange));
```
$$14.57686503$$

To what extent is this result really true? Increase the accuracy.

```
>  evalf(Int(f(x), x = therange),15);
```
$$14.5768650270568$$

2.6 Exercises

E 2.1 Using some properties of measure, namely Proposition 2.1.1 and Proposition 2.1.2, compute the following:

1. the 1-dimensional Lebesgue measure of (a) a one-point set $\{a\}$, $a \in \mathbb{R}$, (b) an interval $[a, b]$, (c) an interval $[a, b)$, (d) the interval $(1, \infty)$.
2. the 2-dimensional Lebesgue measure of (a) the open unit square $(0, 1)^2$, (b) the rectangle $(2, 3) \times [1, 5]$, (c) any segment, (d) any straight line, (e) any given square, (f) any given rectangle, (g) any given triangle.

E 2.2 Find the Lebesgue measure of the set $\mathbb{Z} \subset \mathbb{R}$ of all integers.

E 2.3 Find three different approximation sums for the function $f(x) = \sqrt{1 - x^2}$ defined on the interval $[-1, 1]$ and interpret them graphically.

E 2.4 Find the area of the figure bounded by the graphs of the following functions:
(a) $y = x^2$ and $y = \sqrt{x}$,
(b) $y = \log x$ and $y = 0$,
(c) $y = xe^{-x}$, $y = 0$, when $x \geq 0$,
(d) $y = e^{-x^2}$ and $y = 0$.

E 2.5 Let A be a measurable set of measure zero, i.e. $m(A) = 0$ and let f be an integrable function on A. Prove that $\int_A f\, dm = 0$.

E 2.6 Give an example of an integrable but not summable function defined on (a) the whole real line $\mathbb{R}$, (b) the unit interval $[0, 1]$, (c) the unit square $[0, 1]^2$.

E 2.7 Fix a nonnegative measurable function f and for any measurable set A define

$$\nu(A) = \int_A f \, d\mu. \tag{2.6}$$

Prove that ν is a measure.

Hint. Use Property 8 in Theorem 2.3.1

E 2.8 Find $\nu(A)$ from the previous problem when μ is the 1-dimensional Lebesgue measure, $f(x) = |x|$ and $A = [-2, 3]$.

E 2.9 Let $\{a_i\}$ and $\{m_i\}$, $i = 1, 2, 3, \ldots$ be given sequences of nonnegative real numbers. Show that $\sum_{i=1}^{\infty} a_i m_i$ can be expressed as an integral with respect to the measure on $\mathbb{N}$ defined by $\mu(A) = \sum_{i \in A} m_i$.

Hint. See Example 2.2.3.

E 2.10 Let a_i be given real numbers. Can the series $\sum_{i=1}^{\infty} a_i$ be interpreted as an integral with respect to the counting measure?

Hint. See Example 2.2.3 and use the following MAPLE commands.

```
>   sum((-1)^(i+1)/i,i=1..infinity);
>   sum(1/(2*i),i=1..infinity);
>   sum(1/(2*i+1),i=0..infinity);
>   sum((-1)^(i+1)/i^2,i=1..infinity);
>   evalf(%);
>   sum(1/(2*i+1)^2,i=0..infinity);
>   evalf(%);
>   sum(1/(2*i)^2,i=1..infinity);
>   evalf(%);
>   %%%- %;
```

E 2.11 (a) Express the area of the unit circle as a Riemann integral. (b) Find its exact and approximate value.

E 2.12 Find the area of the unit circle D as the Lebesgue integral $\int_D 1 \, d\mu_2$.

Hint. Change variables using polar coordinates.

E 2.13 Find the area of the figure bounded by the ellipse $4x^2 + y^2 = 1$.

Hint. The ellipse is the image of the unit circle by some linear map.

E 2.14 Find the volume of the unit ball $x^2 + y^2 + z^2 \leq 1$.

E 2.15 Compute the integral $\int_0^3 \frac{\sin x}{x} \, dx$ using various numerical procedures offered by MAPLE. See the Help page int[numeric].

3 Random Variables and Distributions

Most quantities that we deal with every day are more or less of a random nature. The height of the person we first meet on leaving home in the morning, the grade we will earn in the next exam, the cost of a bottle of wine we will buy in the nearest shopping centre, and many more, serve as examples of so called random variables. Any such variable has its own specifications or characteristics. For example, the height of an adult male may take all values from 150 cm to 230 cm or even beyond this interval. Besides, the interval $(175, 180)$ is more likely than other intervals of the same length, say $(152, 157)$ or $(216, 221)$. On the other hand, the grade we will earn during the coming exam will take finite many values: excellent, good, pass, fail, and for the reader some of these values are more likely than others.

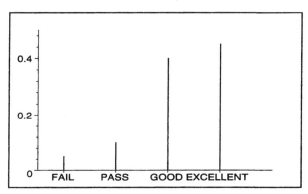

Graphically, the distribution of the grades of the reader may be described by the above figure. The heights of the bars are equal to the corresponding probabilities and their sum equals 1.

For an average student the distribution of his grades might have a different shape.

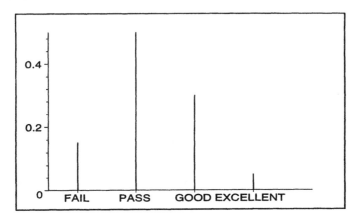

A hypothetical distribution of the height of an adult male may appear as the following picture. The probability here is represented by a corresponding area under the graph. For example, the shaded area in the figure is just the probability that the height of an adult male is greater then 180 cm and less than 185 cm. The whole area under the graph equals 1.

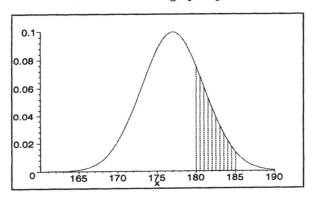

Distributions of random variables can often be characterised by parameters. This makes it possible to compare two or more distributions with each other. Also, the problem of determining and measuring the dependence or correlation of random variables with each other is of great importance. In fact, we believe that the height and weight of a student is more strongly correlated than her height and examination mark.

We will discuss the problems mentioned above in the following sections. First, we will consider probability distributions, then random variables and vectors. In the next chapter basic characteristics of probability distributions and the problem of correlation will be considered.

3.1 Probability Distributions

The real line, real plane, three dimensional space and, generally, the space $\mathbb{R}^n$ are often considered as sets Ω of elementary events. In most cases one considers σ-algebra of all Borel sets $\mathcal{B}(\mathbb{R}^n)$, i.e. the smallest σ-algebra containing all of the open sets. On the other hand, the probability measures P defined on $\mathcal{B}(\mathbb{R}^n)$ can be chosen in a variety of ways.

Definition 3.1.1 *A probability distribution (n-dimensional), briefly a distribution, is a measure P such that the triple*

$$(\mathbb{R}^n, \mathcal{B}(\mathbb{R}^n), P)$$

is a probability space.

Two distinct classes of probability distributions are usually of interest: discrete distributions and continuous distributions. There are also other less typical distributions beyond those contained in these two classes.

3.1.1 Discrete Distributions

Definition 3.1.2 *A distribution $P : \mathcal{B}(\mathbb{R}^n) \longrightarrow \mathbb{R}$ is discrete if there exist sequences $\{x_i\} \subset \mathbb{R}^n$ and $\{p_i\} \subset \mathbb{R}$, $i = 1, 2, \ldots, N$ with $N \leq \infty$ such that:*

$$p_i > 0, \quad \sum_{i=1}^{N} p_i = 1, \quad and \quad P(\{x_i\}) = p_i \quad for\ all\ i = 1, 2, \ldots, N.$$

This definition fully determines the distribution. Namely, for any Borel set A we have:

$$P(A) = P(A \cap \bigcup_{i=1}^{N} \{x_i\}) = \sum_{i=1}^{N} P(A \cap \{x_i\}) = \sum_{i: x_i \in A} P(\{x_i\}).$$

The first equality follows from the fact that $P(\bigcup_{i=1}^{N} \{x_i\}) = 1$ and from Problem 1.6. Thus we have:

$$P(A) = \sum_{i: x_i \in A} p_i. \tag{3.1}$$

The grades of a student just discussed had discrete distributions.

MAPLE ASSISTANCE 19

We graph a discrete one dimensional distribution determined by the sequences: $x_1 = 1$, $x_2 = 2$, $x_3 = 3$, $x_4 = 4$, $x_5 = 5$ and $p_1 = 0.1$, $p_2 = 0.15$, $p_3 = 0.4$, $p_4 = 0.3$, $p_5 = 0.05$.

```
> with(plottools):
> PLOT( CURVES([[1,0],[1,0.1]],[[2,0],[2,0.15]],
  [[3,0],[3,0.4]],[[4,0],[4,0.3] ],[[5,0],[5,0.05]],
  THICKNESS(3)));
```

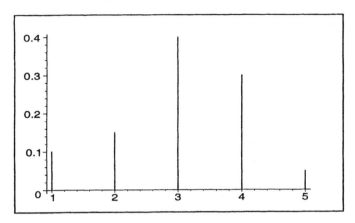

Note carefully that capital letters have been used in the MAPLE command.

MAPLE ASSISTANCE 20

We graph a discrete two dimensional distribution concentrated in points $x_1 = (1,2)$, $x_2 = (1,-1)$, $x_3 = (1,1)$, $x_4 = (0,0)$, $x_5 = (2,0)$, $x_6 = (2,1)$ with corresponding probabilities $p_1 = 0.1$, $p_2 = 0.2$, $p_3 = 0.15$, $p_4 = 0.2$, $p_5 = 0.3$, $p_6 = 0.05$. This time we apply a slightly different approach to that in the previous example.

```
>  x := [1,2],[1,-1],[1,1],[0,0],[2,0],[2,1];
```
$$x := [1, 2], [1, -1], [1, 1], [0, 0], [2, 0], [2, 1]$$
```
>  p := 0.1,0.2,0.15,0.2,0.3,0.05;
```
$$p := .1, .2, .15, .2, .3, .05$$

We make the list of probability bars:
```
>  zip((x,p)->[[op(x),0],[op(x),p]], [x],[p]);
```

$$[[[1, 2, 0], [1, 2, .1]], [[1, -1, 0], [1, -1, .2]], [[1, 1, 0], [1, 1, .15]],$$
$$[[0, 0, 0], [0, 0, .2]], [[2, 0, 0], [2, 0, .3]], [[2, 1, 0], [2, 1, .05]]]$$

```
>  PLOT3D( CURVES(op(%),THICKNESS(5),COLOR(HUE,0.99)),
   AXESSTYLE(BOX),AXESTICKS(3,3,2));
```

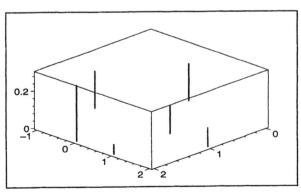

Let us check that it is actually a probability distribution.

```
>  add(p[i],i=1..nops([p]));
```
$$1.00$$

$$\cdots$$

A rather trivial discrete distribution is a one-point distribution, i.e. a distribution P concentrated at a single point, say in point $c \in \mathbb{R}^n$. Then:

$$P(\{c\}) = 1.$$

Despite its simplicity an important discrete distribution is the two-point distribution, i.e. a distribution P concentrated at two points. Usually one considers a one dimensional two-point distribution concentrated in points 0 and 1. Then we have:

$$P(\{0\}) = q, \qquad P(\{1\}) = p,$$

where $p, q > 0$ and $p + q = 1$.

In future, for convenience, we will just write $P(a)$ instead of $P(\{a\})$ for singleton events. Later in this book we will discuss more interesting discrete distributions.

3.1.2 Continuous Distributions

Definition 3.1.3 *A distribution $P : \mathcal{B}(\mathbb{R}^n) \longrightarrow \mathbb{R}$ is called continuous if there exists a function $f : \mathbb{R}^n \longrightarrow \mathbb{R}$, called its density function, such that for any Borel set $A \subset \mathbb{R}^n$ we have:*

$$P(A) = \int_A f(x)\, dx = \int_A f\, dx, \qquad (3.2)$$

where the integral is taken with respect to the Lebesgue measure $\mu = \mu_n$ restricted to $\mathcal{B}(\mathbb{R}^n)$ (see Section 2.2).

An example of a continuous distribution is height of a man shown in the figure on page 62. The probability of any set A equals the area under the graph over A.

Note that a density function f has to be integrable and satisfy the two following conditions:

1. $f \geq 0$ a.e.
2. $\int_{\mathbb{R}^n} f(x)\, dx = 1$

One can show that an integrable function f satisfying these two conditions is the density function of some continuous distribution (see E 2.7.

Later in the book we will examine some basic continuous distributions. For now we consider the simplest such distribution.

Example 3.1.1 Let G be a Borel set of positive and finite Lebesgue measure. Define a function

$$f(x) = \begin{cases} 0, & \text{if } x \notin G \\ \dfrac{1}{\mu(G)} & \text{if } x \in G \end{cases}$$

This f clearly satisfies the conditions of a density function. The resulting distribution is called the *uniform distribution* on the set G, and sometimes is denoted by $U(G)$. Compare this example with the definition of the geometric probability (see Definition 2.1.3).

Let us graph the density of the uniform distribution on the interval $(2, 4)$, $U(2, 4)$.

MAPLE ASSISTANCE 21

```
>  f := x->piecewise(x>2 and x<4,1/2):
>  plot(f,0..5,discont=true,thickness=3,color=BLACK);
```

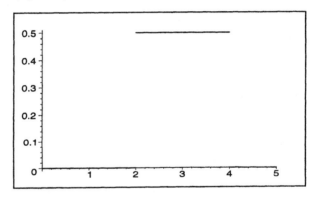

As we have already mentioned in the case of a continuous one dimensional distribution, it is easy to "see" the probability of a given set. For an n-dimensional distribution with density f it is the Lebesgue measure of the set:

$$\{(x, y) \in \mathbb{R}^n \times \mathbb{R} : x \in A, 0 \leq y \leq f(x)\}.$$

This indicates that the probability of all singletons and thus that of any finite or countable set is zero. In fact, formula (3.2) applied to a singleton set A immediately implies this assertion.

3.1.3 Distribution Function

One dimensional probability distributions play an exceptional role amongst probability distributions. Here the use of distribution functions greatly simplifies the examination of distributions.

Definition 3.1.4 *A function $F : \mathbb{R} \longrightarrow \mathbb{R}$ is called a distribution function or a cumulative distribution function if the following conditions are satisfied:*

1. *F is increasing: $x < y \Rightarrow F(x) \leq F(y)$;*
2. *F is continuous from the right: $\lim_{x \to a^+} F(x) = F(a)$, for all $a \in \mathbb{R}$;*
3. *$\lim_{x \to \infty} F(x) = 1$;*
4. *$\lim_{x \to -\infty} F(x) = 0$.*

There is a close correspondence between probability distributions and distribution functions.

Theorem 3.1.1 *If P is a one dimensional probability distribution, then the function $F = F_P$ defined by*

$$F(x) = P((-\infty, x]) = P(-\infty, x] \qquad (3.3)$$

is a distribution function. ($P((-\infty, x])$ will usually be written $P(-\infty, x]$).

Conversely: for any distribution function F there exists a unique probability distribution P such that formula (3.3) is satisfied.

Proof. Let P be a probability distribution.

1. For $x < y$ we have: $(-\infty, x] \subset (-\infty, y]$, so $F(x) = P(-\infty, x] \leq P(-\infty, y] = F(y)$.

2. Let $x_n \to a^+$, as $n \to \infty$, and assume it is a decreasing sequence (i.e. $x_n > x_{n+1} > a$). To prove Point 2 we will show that $F(x_n) \to F(a)$. We have $F(x_n) = P(-\infty, x_n]$, and the sets $(-\infty, x_n]$ form a decreasing sequence with the intersection $= (-\infty, a]$. Property 7 in Proposition 1.1.1 completes the proof.

3. The arguments are quite similar to those above. We just have to note that for an increasing sequence with x_n going to infinity, the sets $(-\infty, x_n]$ form an increasing sequence with their union equal to $\mathbb{R}$.

4. As in Point 3, but we take a decreasing sequence x_n tending to $-\infty$.

The proof of the second statement is more advanced and therefore will be omitted. Readers interested in the proof are referred to Halmos [13].   □

Incidentally, some authors define a distribution function with condition 2 replaced by the condition that F is continuous from the left. In such a case formula (3.3) becomes $F(x) = P(-\infty, x)$. Certainly, both approaches are equally valid.

· · ·

An important question concerns the nature of a discontinuity of a distribution function at a given point. The following proposition brings the answer.

Proposition 3.1.1 *Let P be a probability distribution, $F = F_P$ be a corresponding distribution function, and $a \in \mathbb{R}$. Then*

$$P(a) = F(a) - F(a^-), \tag{3.4}$$

where $F(a^-)$ denotes the left hand side limit of F at the point a. (This limit exists since F is increasing).

Hence F is continuous at a if and only if $P(a) = 0$.

Proof. Let x_n be an increasing sequence. Then $(-\infty, a) = \bigcup_n (-\infty, x_n]$ and hence $F(a^-) = \lim_{n \to \infty} F(x_n) = \lim_{n \to \infty} P(-\infty, x_n] = P(-\infty, a)$. Hence $P(a) = P((-\infty, a] \backslash (-\infty, a)) = P(-\infty, a] - P(-\infty, a) = F(a) - F(a^-)$.
□

In other words, a continuity point of a distribution function has zero probability, and a discontinuity point has probability equal to the "jump" of the distribution function.

. . .

The distribution function is easy to find for both discrete and continuous probability distribution. We have:

1. In the discrete case:

$$F_P(x) = \sum_{i : x_i \le x} p_i. \tag{3.5}$$

2. In the continuous case:

$$F_P(x) = \int_{-\infty}^{x} f(s) \, ds. \tag{3.6}$$

The distribution function of a continuous probability distribution is thus continuous at every point. The converse is not true, however. Moreover, (3.6) implies that

$$\frac{d}{dx} F_P(x) = f(x) \tag{3.7}$$

at any continuity point x of the density f.

Example 3.1.2 The distribution function of the uniform distribution $U(a, b)$ on the interval (a, b) is

$$F(x) = \begin{cases} 0, & x < a \\ \dfrac{x - a}{b - a}, & a \le x < b \\ 1, & b \le x \end{cases}$$

We graph the distribution function of the uniform distribution on the interval $(2,4)$.

MAPLE ASSISTANCE 22

```
> F := x->piecewise(x>2 and x<4,(x-2)/2,x>4,1):
> plot(F,0..5,thickness=3,color=BLACK);
```

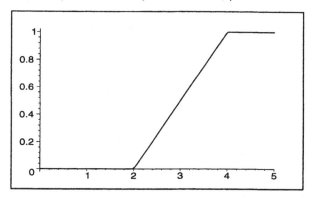

We check the validity of formula (3.7) in the above case.

MAPLE ASSISTANCE 23

```
> diff(F(x),x);
```

$$
\begin{cases}
0 & x < 2 \\
undefined & x = 2 \\
\frac{1}{2} & x < 4 \\
undefined & x = 4 \\
0 & 4 < x
\end{cases}
$$

$\cdots$

The definition and some properties of a distribution function can also be established in n-dimensional space with a similar correspondence between distribution functions and probability distributions to that in the one dimensional case. The only two complications are with an appropriate formulation of properties that in one dimension are expressed by means of the ordering "$\leq$" of the real line and by means of ∞ and $-\infty$.

3.2 Random Variables and Random Vectors

We will start with the definition of a random variable and then generalise it to the definition of a random vector. Let (Ω, Σ, P) be a probability space.

Definition 3.2.1 *A function* $\xi : \Omega \longrightarrow \mathbb{R}$ *is called a random variable if it is measurable, i.e.:*

$$
\xi^{-1}(B) = \{\omega \in \Omega : \xi(\omega) \in B\} \in \Sigma,
$$

for any set $B \in \mathcal{B}(\mathbb{R})$.

Sets of the form $\xi^{-1}(B)$, where $B \in \mathcal{B}(\mathbb{R})$, are said to be specified by ξ. We will also use the shorter notation $\{\xi \in B\}$ for $\xi^{-1}(B)$. For example: $P(\xi < \varepsilon)$ means $P(\{\omega \in \Omega : \xi(\omega) < \varepsilon\})$.

3.2.1 A Problem of Measurability

We would like to explain the idea of the above definition, especially since many applied statistics texts omit or postpone the assumption on measurability and call any function $\xi : \Omega \longrightarrow \mathbb{R}$ a random variable. This may not cause difficulties in some simple cases, but may lead to serious mistakes in more advanced cases. We can think of measurability as being consistency with the information in the probability space under consideration.

Let us note first that in the case $\Sigma = \mathcal{P}(\Omega)$, as in the classical scheme $\xi^{-1}(B) \in \Sigma$ for any B, any function $\xi : \Omega \longrightarrow \mathbb{R}$ is a random variable. On the other hand, if Σ is smaller than $\mathcal{P}(\Omega)$, then one can find functions $\xi : \Omega \longrightarrow \mathbb{R}$ that are not random variables. In the extreme case, where Σ is the σ-algebra containing only the empty set $\emptyset$ and the whole space Ω, only the constant functions $\Omega \longrightarrow \mathbb{R}$ are random variables. The following example illustrates what is and what is not a random variable.

Example 3.2.1 Consider an experiment of tossing two dice where we are interested in the maximum of pips. The set of elementary events is then $\Omega = \{(i,j) : i,j = 1,\ldots,6\}$, the σ-algebra $\Sigma = \sigma(A_1, \ldots A_6)$ is the σ-algebra of all unions of sets $A_k = \{(i,j) : \max(i,j) = k\}$. (The measure P here may be defined quite arbitrarily.) Define three functions $\Omega \longrightarrow \mathbb{R}$:

$$\xi_1(i,j) = \max(i,j),$$

$$\xi_2(i,j) = \begin{cases} 1, & \text{if } i > 3 \text{ or } j > 3, \\ 0, & \text{otherwise,} \end{cases}$$

$$\xi_3(i,j) = i + j.$$

It is easy to see that the first and the second function are random variables on (Ω, Σ, P), while the third function is not. In fact, the set $\xi_3^{-1}(-\infty, 3] = \{(1,1), (1,2), (2,1)\}$ is not a union of sets A_k, so does not belong to Σ.

The idea here is that the sets specified by the first two functions provide some information about the experiment. For example $\xi_2 = 0$ means the maximum ≤ 3, and therefore they are events. On the other hand, sets specified by ξ_3 may not be events and we are not able to say here whether or not $\xi_3^{-1}(-\infty, 3] = \{(1,1), (1,2), (2,1)\}$ occurs.

3.2.2 Distribution of a Random Variable

A random variable induces a probability distribution on the measurable space $(\mathbb{R}, \mathcal{B}(\mathbb{R}))$, which is often more convenient to work with.

Definition 3.2.2 *Let* (Ω, Σ, P) *be a probability space and let* $\xi : \Omega \longrightarrow \mathbb{R}$ *be a random variable. The probability distribution* P_ξ *defined by*

$$P_\xi(B) = P(\xi^{-1}(B)), \text{ for } B \in \mathcal{B}(\mathbb{R}) \tag{3.8}$$

is called the distribution induced by the random variable ξ

We note that the measurability of ξ guarantees the validity of the expression $P(\xi^{-1}(B))$. Also, one easily checks that P_ξ is in fact the probability measure. Thus $(\mathbb{R}, \mathcal{B}(\mathbb{R}), P_\xi)$ is a probability space.

Example 3.2.2 Find the distribution of the random variable ξ_2 from Example 3.2.1. Assume that the measure P is defined according to the classical scheme. We have: $P_\xi(x) = P(\xi^{-1}(x)) = 0$ for $x \neq 0$ and $x \neq 1$, and $P_\xi(0) = P(A_1 \cup A_2 \cup A_3) = \frac{9}{36}$, $P_\xi(1) = P(A_4 \cup A_5 \cup A_6) = \frac{27}{36}$. So P_ξ is a two-point distribution.

Some examples of random variables and their distributions have already been mentioned on page 61. However, we said nothing about the probability space the random variables were defined on. Actually, it is a typical situation. Most often we deal with a random quantity and what we really see is a probability distribution characterising this quantity. Then we believe that the quantity is a random variable ξ on some probability space (Ω, Σ, P) and the probability distribution we see is just P_ξ. This, perhaps rather naive, reasoning is fortunately justified by the following:

Theorem 3.2.1 *Let* $Q : \mathcal{B}(\mathbb{R}) \longrightarrow \mathbb{R}$ *be a probability distribution. Then there is a probability space* (Ω, Σ, P) *and a random variable* $\xi : \Omega \longrightarrow \mathbb{R}$ *such that*

$$Q = P_\xi.$$

Proof. Simply take $\Omega = \mathbb{R}$, $\Sigma = \mathcal{B}(\mathbb{R})$, $P = Q$ and $\xi(\omega) = \omega$ for $\omega \in \mathbb{R}$. $\quad\square$

$\cdots$

Random variables share a lot of useful algebraic properties with general measurable functions (see Theorem 2.2.2 on page 43). We just recall that:

1. The indicator I_A of a set $A \in \Sigma$ is a random variable.
2. The sum, difference, product, quotient (if defined) of random variables is a random variable.
3. The minimum and maximum of two random variables is a random variable.

4. The supremum and infimum of a sequence of random variables is a random variable.
5. The limit of a sequence of random variables is a random variable.
6. The composition $g \circ \xi$, of a random variable ξ with a continuous function g is a random variable.

3.2.3 Random Vectors

A random vector is a straightforward generalisation of a random variable.

Definition 3.2.3 *A function $\xi : \Omega \longrightarrow \mathbb{R}^n$ is called a random vector if it is measurable, i.e.*

$$\xi^{-1}(B) = \{\omega \in \Omega : \xi(\omega) \in B\} \in \Sigma, \quad \textit{for any set } B \in \mathcal{B}(\mathbb{R}^n).$$

A random variable is a one dimensional random vector, so there were no formal reasons to distinguish the random variables case. We did so, however, due to tradition and because of the special role random variables play among random vectors.

As with random variables we can define the distribution of a random vector $X : \Omega \longrightarrow \mathbb{R}^n$ by:

$$P_X(B) = P(X^{-1}(B)), \text{ for } B \in \mathcal{B}(\mathbb{R}^n). \tag{3.9}$$

3.3 Independence

A pair (X, Y) of random vectors X, Y defined on the same probability space, in particular a pair of random variables, is a random vector. What is the relationship between the distributions P_X, P_Y and $P_{(X,Y)}$? In general, there is only a partial relationship, namely, the definition of the distribution of the random vector implies:

$$P_X(A_1) = P_{(X,Y)}(A_1 \times \mathbb{R}^n), \qquad P_Y(A_2) = P_{(X,Y)}(\mathbb{R}^m \times A_2). \tag{3.10}$$

Hence, the knowledge of the distribution of the pair of random vectors implies the knowledge of the distributions of its "coordinates". Note that knowing the probability space that the random vectors are defined on is not essential here. On the other hand, the probability space is important for questions like the converse problem: determine the distribution $P_{(X,Y)}$ from the distributions P_X and P_Y. The problem is illustrated with the following:

Example 3.3.1 Consider two different situations.
   Case 1. Let (Ω, Σ, P) be a classical scheme with $\Omega = \{1, 2, 3, 4, 5, 6\}$, $\xi, \eta : \Omega \longrightarrow \mathbb{R}$ defined as: $\xi(\omega) = \eta(\omega) = \omega$. Here the measures $P_\xi = P_\eta$ are concentrated at the points $1, 2, 3, 4, 5, 6$ with equal probabilities $\frac{1}{6}$, while the

measure $P_{(\xi,\eta)}$ is concentrated at the points $x_i = (i,i)$, $i = 1,\ldots,6$ with $P_{(\xi,\eta)}(x_i) = \frac{1}{6}$.

Case 2. Let (Ω, Σ, P) be the classical scheme with $\Omega = \{1,2,3,4,5,6\}^2$, $\xi, \eta : \Omega \longrightarrow \mathbb{R}$ defined as: $\xi(\omega_1, \omega_2) = \omega_1$, $\eta(\omega_1, \omega_2) = \omega_2$. As above $P_\xi = P_\eta$ are concentrated at the points $1,2,3,4,5,6$ with equal probabilities $\frac{1}{6}$, while the measure $P_{(\xi,\eta)}$ is concentrated at the points $x_{ij} = (i,j)$, $i,j = 1,\ldots,6$ with $P_{(\xi,\eta)}(x_{ij}) = \frac{1}{36}$.

In both cases the random variables ξ, η had the same distributions, but the distributions of the pair (ξ, η) were different.

Note that in Case 2 we have: $P_{(\xi,\eta)}(i,j) = P_\xi(i)P_\eta(j)$ for all i,j, which means that the distribution $P_{(\xi,\eta)}$ is the (Cartesian) product of P_ξ and P_η. The fundamental reason is that the random variables in Case 2 are independent. What independence of the random variables means will be formally defined below, though we already believe that the toss of a die is independent of the toss of another die, so ξ, η "should be" independent.

Definition 3.3.1 *Let (Ω, Σ, P) be a probability space and let $X_1, \ldots, X_k$ be random vectors defined on this space. We say that $X_1, \ldots, X_k$ are independent if for every choice of Borel sets $B_1, \ldots, B_k$ contained in the appropriate spaces we have:*

$$P(X_1 \in B_1, \ldots, X_k \in B_k) = P(X_1 \in B_1) \cdot \ldots \cdot P(X_k \in B_k). \qquad (3.11)$$

In other words, random vectors are independent if the events specified by these vectors are independent.

We say that the random vectors $X_1, X_2, X_3, \ldots$ are independent, if for any k, the random vectors $X_1, \ldots, X_k$ are independent.

In the last example it is obvious that we have a pair of dependent random variables in Case 1 and a pair of independent random variables in Case 2. This example suggests the following:

Theorem 3.3.1 *Let X, Y be random vectors defined on the same probability space (Ω, Σ, P). Then*

$$X, Y \text{ are independent} \iff P_{(X,Y)} = P_X \times P_Y.$$

Moreover, we then have:

1. *If random vectors X and Y have discrete distributions and*

$$P(X = x_i) = p_i, \qquad P(Y = y_j) = q_j,$$

then (X,Y) has a discrete distribution with

$$P(X = x_i, Y = y_j) = p_i \cdot q_j \qquad (3.12)$$

2. *If random vectors X and Y have continuous distributions with densities f and g, then (X,Y) has the continuous distribution with density*

$$h(x,y) = f(x) \cdot g(y) \qquad (3.13)$$

The above formulas are very useful for determining distributions of functions of random variables and vectors and will be used later in the book.

3.4 Functions of Random Variables and Vectors

The following situation is often encountered. Given a random vector X with known distribution P_X and given a function φ, find the distribution $P_{\varphi(X)}$ of the random vector $\varphi(X) = \varphi \circ X$. For example, we may want to find the distribution of ξ^2 given the distribution of a random variable ξ, or we may want find the distribution of $\max(\xi, \eta)$ given the distributions of ξ and η.

The theoretical solution to this problem is remarkably easy, but its practical meaning is fairly limited. Nevertheless, we present it first and will then consider a more practical approach.

Let (Ω, Σ, P) be a probability space, $X : \Omega : \longrightarrow \mathbb{R}^n$ a random vector and $\varphi : \mathbb{R}^n : \longrightarrow \mathbb{R}^k$ a function. We are interested in the composition

$$\varphi \circ X : \Omega \longrightarrow \mathbb{R}^k \quad \text{defined by} \quad (\varphi \circ X)(\omega) = \varphi(X(\omega)).$$

For a broad class of functions φ one can prove that this composition is a random vector, for example, if φ is a Borel function, i.e. measurable with respect to the space $(\mathbb{R}^n, \mathcal{B}(\mathbb{R}^n))$. For historical reasons we often write $\varphi(X)$ instead of $\varphi \circ X$. Its distribution is then:

$$P_{\varphi(X)}(C) = P((\varphi \circ X)^{-1}(C)) = P(X^{-1}(\varphi^{-1}(C))) = P_X(\varphi^{-1}(C))$$

for any Borel subset C of the space $\mathbb{R}^k$. As we have already mentioned this nice and general formula is not very convenient in practice because one usually considers distribution functions or densities rather than probability distributions.

In the following we offer a few examples illustrating common methods of finding distributions of functions of random vectors.

Example 3.4.1 Let ξ be a random variable having discrete distribution concentrated at the points $-1, 0, 1, 2, 3$ with equal probability $\frac{1}{5}$. The distribution of the random variable ξ^2 is found as follows: ξ^2 takes values $0, 1, 4, 9$ with probabilities, respectively, $\frac{1}{5}, \frac{2}{5}, \frac{1}{5}, \frac{1}{5}$, since $(\pm 1)^2 = 1$.

Example 3.4.2 Let ξ be a random variable, $F = F_\xi$ its distribution function and let $a, b \in \mathbb{R}$ be fixed numbers with $a \neq 0$. We seek the distribution function of the random variable $\eta = a\xi + b$.

For $a > 0$ we have

$$F_\eta(x) = P(\eta \leq x) = P(a\xi + b \leq x) = P\left(\xi \leq \frac{x-b}{a}\right) = F_\xi\left(\frac{x-b}{a}\right).$$

Similarly, for $a < 0$ we have

$$F_\eta(x) = P(\eta \le x) = P(a\xi + b \le x) = P\left(\xi \ge \frac{x-b}{a}\right)$$

$$= 1 - P\left(\xi < \frac{x-b}{a}\right) = 1 - F_\xi\left(\left(\frac{x-b}{a}\right)^-\right).$$

Assume additionally that ξ has a density f, where f is a continuous function. This is not necessary as the next example shows, but simplifies arguments here. Then (3.7) means that F_ξ is differentiable, so is F_η from the above formula too. By (3.7) η has a continuous distribution with density:

$$g(x) = \frac{1}{|a|} f\left(\frac{x-b}{a}\right). \tag{3.14}$$

Example 3.4.3 We generalise Example 3.4.2 to some extent. Let X be a random vector having an n-dimensional continuous probability distribution with density $f : \mathbb{R}^n \longrightarrow \mathbb{R}$. Assume $\varphi : \mathbb{R}^n \longrightarrow \mathbb{R}^n$ is a diffeomorphism, i.e. φ is invertible with both φ and φ^{-1} having continuous first derivatives. Then, the random vector $\varphi(X)$ also has continuous distribution with density g given by:

$$g(x) = \left| Jac_x \varphi^{-1} \right| f(\varphi^{-1}(x)),$$

where the Jacobian $Jac_x \varphi = \det \left[\dfrac{\partial \varphi_i}{\partial x_j} \right]$ is nonzero everywhere.

In fact, by Theorem 2.3.3, for any Borel set A we have

$$P(\varphi(X) \in A) = P(X \in \varphi^{-1}(A))$$

$$= \int_{\varphi^{-1}(A)} f(y)\, dy = \int_A f(\varphi^{-1}(x)) \left| Jac_x \varphi^{-1} \right| dx.$$

Example 3.4.4 Let ξ be a random variable having uniform distribution on the interval $[-1, 1]$, i.e. its density is $f(x) = \frac{1}{2} I_{[-1,1]}$, where $I_{[-1,1]}$ is the indicator function of the interval $[-1, 1]$. We determine the distribution of the random variable $\frac{1}{\xi}$ by computing its distribution function G.

We have to compute $G(a)$ for any $a \in \mathbb{R}$. Let $a < 0$. Then

$$G(a) = P\left(\frac{1}{\xi} \le a\right) = P\left(\frac{1}{\xi} \le a, \xi < 0\right) = P\left(\frac{1}{a} \le \xi < 0\right)$$

$$= \int_{\frac{1}{a}}^0 f(x)\, dx = \begin{cases} -\dfrac{1}{2a}, & \text{for } a \le -1 \\[2mm] \dfrac{1}{2}, & \text{for } -1 < a \le 0. \end{cases}$$

The cases $a > 0$ and $a = 0$ are even simpler and hence are left to the reader. We then have

$$G(a) = \begin{cases} -\dfrac{1}{2a}, & \text{for } a \leq -1 \\ \dfrac{1}{2}, & \text{for } -1 < a \leq 1 \\ 1 - \dfrac{1}{2a}, & \text{for } 1 < a. \end{cases}$$

We see that the distribution function G is differentiable at any point except at -1 and 1. Formula (3.7) can be used to determine the density:

$$g(a) = G'(a) = \begin{cases} \dfrac{1}{2a^2}, & \text{for } a < -1 \\ 0, & \text{for } -1 < a < 1 \\ \dfrac{1}{2a^2}, & \text{for } 1 < a. \end{cases}$$

We sketch the graph of the distribution function G and the density g:

MAPLE ASSISTANCE 24

```
>  G := a->piecewise(a<=-1,-1/(2*a),
   -1<a and a <=1,1/2,
   1<a,1-1/(2*a));
```

$$G := a \rightarrow \text{piecewise}(a \leq -1, -\frac{1}{2}\frac{1}{a}, -1 < a \text{ and } a \leq 1, \frac{1}{2}, 1 < a, 1 - \frac{1}{2}\frac{1}{a})$$

```
>  plot(G);
```

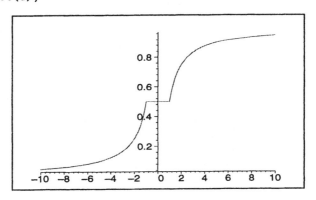

```
>  g := D(G);
```

$$g := a \rightarrow \text{piecewise}(a \leq -1, \frac{1}{2}\frac{1}{a^2}, -1 < a \text{ and } a \leq 1, 0, 1 < a, \frac{1}{2}\frac{1}{a^2})$$

```
>  plot(g,discont=true,thickness=3,color=BLACK);
```

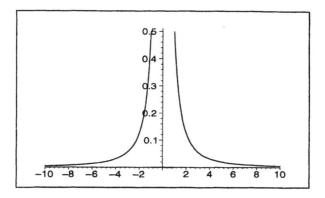

3.4.1 Distributions of a Sum

Example 3.4.5 We will discuss the distribution of the sum of a random variable for the continuous case only since the discrete one is simpler and is thus left to the reader (see Problems 3.10 and 3.11).

Assume that a two dimensional random vector (ξ, η) has continuous distribution with density f. First we find the distribution function of the sum $F = F_{\xi+\eta}$. Fix $a \in \mathbb{R}$. Then, $F(a) = P(\xi + \eta \leq a) = P((\xi, \eta) \in A)$, where $A = \{(x, y) : x + y \leq a\}$. Hence,

$$F(a) = \int_A f(x, y)\, d(x, y) = \int_{-\infty}^{a} \left(\int_{-\infty}^{\infty} f(t, s - t)\, dt \right)\, ds.$$

We have used the substitution $t = x$, $s = x + y$. Differentiating with respect to a and setting $x = a$ we obtain the formula for the density of the sum:

$$f_{\xi+\eta}(x) = \int_{-\infty}^{\infty} f(t, x - t)\, dt.$$

Assume now that the random variables ξ and η are independent. Then, by formula (3.13), the last equality simplifies to:

$$f_{\xi+\eta}(x) = \int_{-\infty}^{\infty} f_\xi(t) f_\eta(x - t)\, dt. \tag{3.15}$$

$\cdots$

To complete this section we state a useful theorem.

Theorem 3.4.1 *Let $X : \Omega \longrightarrow \mathbb{R}^n$ and $Y : \Omega \longrightarrow \mathbb{R}^m$ be independent random vectors and let $g : \mathbb{R}^n \longrightarrow \mathbb{R}^k$, $h : \mathbb{R}^m \longrightarrow \mathbb{R}^l$ be Borel functions, i.e. measurable with respect to appropriate σ-algebras of Borel sets. Then the random vectors $g(X)$ and $h(Y)$ are independent.*

3.5 MAPLE Session

M 3.1 *Write* MAPLE *commands for the figure on page 62 of the distribution of height.*

```
>   sigma := 4: m := 177:
>   w := 1/(sqrt(2*Pi)*sigma);
```
$$w := \frac{1}{8} \frac{\sqrt{2}}{\sqrt{\pi}}$$
```
>   f := x -> w*exp((-((x-m)/sigma)^2/2));
```
$$f := x \rightarrow w\, e^{(-1/2\,\frac{(x-m)^2}{\sigma^2})}$$
```
>   with(plottools):
>   an := plot(f(x),x = 160..190):
    bn := seq(curve([[i/2,0],[i/2,f(i/2)]],linestyle = 2),
    i = 360..370):
>   plots[display]({an,bn},axes=FRAMED,
    view=[160..190,0.. 0.1]);
```

M 3.2 *Verify that the function f used in the previous problem really is a density function.*

```
>   int(f,-infinity..infinity);
                    1
```

M 3.3 *Write* MAPLE *commands for the figure on page 61.*

```
>   PLOT(CURVES([[2,0],[2,0.05]],[[3,0],[3,0.1]],
    [[4,0],[4,0.4]],[[5,0],[5,0.45]],THICKNESS(3)),
    AXESTICKS([2='NOPASS',3='PASS',4='GOOD',5='EXCELLENT'],2),
    VIEW(1.5..6,0..0.5));
```

M 3.4 *Write a procedure that determines the distribution function for any finite sequences* $\{x_i\}$ *and* $\{p_i\}$ *defining a discrete distribution.*

```
>   FF := proc(x::list,p::list,xi)
    f[0] := 0:
    for i from 1 to nops(x) do
    f[i] := f[i-1] + p[i]
    od:
    if xi < x[1] then 0
    elif xi >= x[nops(x)] then 1
    else
    for i from 1 to nops(x)-1 do
    if x[i] <= xi and xi < x[i+1] then RETURN(f[i]) fi
    od
    fi
    end:
```
Warning, 'f' is implicitly declared local

Warning, 'i' is implicitly declared local

Specify some sequences:
```
>  x := [1,3,4,6]: p := [0.2,0.3,0.4,0.1]:
```

Define the distribution:
```
>  F := xi->FF(x,p,xi);
```
$$F := \xi \to FF(x, p, \xi)$$

Does it work?
```
>  F(3);
```
$$.5$$

Make the plot:
```
>  plot(F,-3..8,thickness=3,color=BLACK);
```

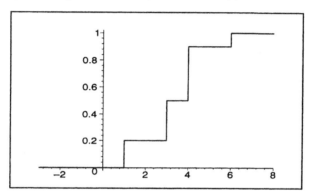

The above method works under an assumption that the sequence x is given in the natural order. If not, then we have to order it first, remembering that the correspondence between the x_i's and p_i's is essential.

Fix some sequences:
```
>  x := [1,4,3,5,2]: p := [0.05,0.3,0.4,0.1,0.05]:
```

Define the criterion function which is necessary to indicate the order of pairs by their first elements.
```
>  crit := (w,z)->(w[1] <= z[1]);
```
$$crit := (w, z) \to w_1 \le z_1$$

Combine the lists x and p into a list of pairs:
```
>  a := [seq([x[i],p[i]],i=1..nops(x))];
```
$$a := [[1, .05], [4, .3], [3, .4], [5, .1], [2, .05]]$$

Sort the list a by applying crit. Pay particular attention to the use of the composition operator @ and the command is.

```
>  a := sort(a,is@crit);
```
$$a := [[1, .05], [2, .05], [3, .4], [4, .3], [5, .1]]$$

Return now to the lists x and p.

```
>  x := [seq(a[i][1],i=1..nops(x))];
```
$$x := [1, 2, 3, 4, 5]$$
```
>  p := [seq(a[i][2],i=1..nops(x))];
```
$$p := [.05, .05, .4, .3, .1]$$

Plot the distribution removing unwanted vertical segments.

```
>  plot(F,-1..8,discont=true,thickness=3,color=BLACK);
```

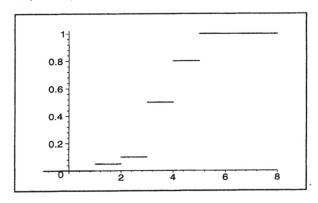

M 3.5 *Plot the distribution function of the random variable taking points* $1, 2, 3, 4$ *with probabilities* $0.2, 0.3, 0.4, 0.1$.

This time we will use the stats package.

```
>  restart:
>  with(stats):
>  F := x ->
     statevalf[dcdf,empirical[0.2,0.3,0.4,0.1]](floor(x));
```
$$F := x \to statevalf_{dcdf, empirical_{.2, .3, .4, .1}}(floor(x))$$

```
> plot(F,-1..6,thickness=2);
```

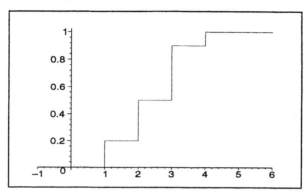

M 3.6 *For any given $p \in [0,1]$ define the inverse distribution function by $invF(p) = \inf\{x \in \mathbb{R} : p \leq F(x)\}$. Plot $invF(p)$ for a previous example.*

```
> invF := x->statevalf[idcdf,empirical[0.2,0.3,0.4,0.1]](x);
> plot(invF,-0.5..1.5,thickness=2);
```

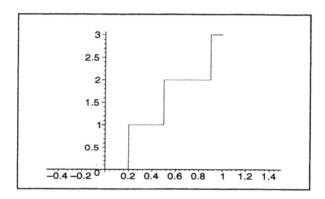

M 3.7 *Find the densities of sums $S_n = \xi_1 + \ldots + \xi_n$ for $n = 2,\ldots,7$, when $\xi_1,\ldots,\xi_n$ are independent random variables with the common density*

$$f(x) = \begin{cases} 0, & x < 0 \\ 3e^{-3x}, & 0 \leq x. \end{cases}$$

Define the convolution using formula (3.15). Note that for a function that is zero for $x < 0$ the limits of integration are in fact 0 and x in formula (3.15).

```
> conv := proc(f,g);
  x -> int(f(x-t)*g(t),t=0..x);
  end;
```
$$conv := \mathbf{proc}(f, g) \, x \to \text{int}(f(x - t) \times g(t), t = 0..x) \, \mathbf{end}$$

Define the required density.

```
>  f := x -> lambda*exp(-lambda*x);
```
$$f := x \rightarrow \lambda e^{(-\lambda x)}$$

Note that $S_{i+1} = S_i + \xi_{i+1}$ and use the result from Example 3.4.5.

```
>  h1 := f:
>  for i from 1 to 6 do
   conv(h.i,f)(x):
>  h.(i+1) := unapply(%,x)
   od:
```

Put $\lambda = 3$. Check that $h7$ really is a density.

```
>  lambda := 3:
>  int(h.7,0..infinity);
```
$$1$$

Plot all the densities on the same axes:

```
>  plot({h.($2..7)},0..5,color=BLACK);
```

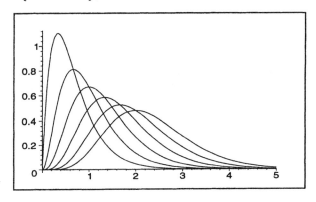

3.6 Exercises

E 3.1 Find a constant C so that the function $P : \mathcal{B}(\mathbb{R}) \longrightarrow \mathbb{R}$ taking values $P(k) = \frac{C}{k(k+1)}$ for $k = 1, 2, 3, \ldots$, is a distribution.

E 3.2 Find a constant C so that the function

$$f(x) = \begin{cases} 0, & \text{for } x < 1 \\ \dfrac{C}{x^4}, & \text{for } x \geq 1 \end{cases}$$

is a density of a probability distribution P and determine $P(\{x \in \mathbb{R} : \frac{1}{x} < a\})$ for a given a.

E 3.3 Given a distribution P on $\mathbb{R}^2$ define the marginal distributions:

$$P_1(A) = P(A \times \mathbb{R}) \quad \text{and} \quad P_2(B) = P(\mathbb{R} \times B),$$

where A and B are Borel sets. Is it true that $P = P_1 \times P_2$?

E 3.4 Find $F(0)$, $F(0.3)$ and $F(3)$ for a distribution function F defined by $F = \frac{1}{3}F_1 + \frac{2}{3}F_2$, where F_1 is the distribution function of the discrete distribution with common value $\frac{1}{5}$ at points $0, 1, 2, 3, 4$, and F_2 is the distribution function of the uniform distribution on the interval $(0, 2)$. Sketch the graph F. Does F correspond to a continuous or discrete distribution?

Hint. Use the Law of Total Probability.

E 3.5 Let $\xi : \Omega \longrightarrow \mathbb{R}$ and assume that Ω is a union of pairwise disjoint sets $A_1, A_2, \ldots, A_k$. Let Σ be the σ-algebra consisting of all unions formed by these sets. Prove that ξ is Σ-measurable if and only if ξ is constant on each of the sets A_i, $i = 1, \ldots, k$.

E 3.6 Find $P(|\xi| > a)$, for $a = 0, \frac{1}{2}, 1, 2$, with the distribution F of a random variable ξ given by:

$$F(x) = \begin{cases} 0, & \text{for } x < -1 \\ \dfrac{1}{3}, & \text{for } -1 \le x < 0 \\ \dfrac{1}{3}(x+1), & \text{for } 0 \le x < 1 \\ 1, & \text{for } 1 \le x. \end{cases}$$

E 3.7 We choose an integer a randomly from the interval $[1, 10]$ according to the classical scheme and then we choose an integer ξ randomly from the interval $[1, a]$, again according to the classical scheme. Find the distribution of ξ. Determine two probability spaces on which the random variable ξ could be defined.

E 3.8 The independent random variables ξ and η are uniformly distributed on the interval $[0.1]$. Find the distributions of $\min(\xi, \eta)$ and $\max(\xi, \eta)$. Are these two variables independent?

E 3.9 Find the distribution of the sum of independent random variables that are uniformly distributed on the interval $[0, 1]$.

E 3.10 Let independent variables ξ and η have discrete distributions $P(\xi = k) = p_k$, $P(\eta = k) = q_k$, $k = 0, 1, 2, 3, \ldots$. Prove that the sum $\xi + \eta$ has the distribution

$$P(\xi + \eta = k) = \sum_{i=0}^{k} p_i q_{k-i}, \quad k = 0, 1, 2, 3, \ldots \tag{3.16}$$

E 3.11 Derive a formula like (3.16) assuming that both variables take all their values in a set $\{1, \ldots, N\}$, where N is a fixed integer.

E 3.12 Does the independence of ξ and η imply the independence of $\xi + \eta$ and $\xi - \eta$?

E 3.13 There are 5 white balls and 3 black balls in a box. Let T_w and T_b denote the number of drawings in which we first get, respectively, a white ball or a black ball. Find (a) The distributions of T_w and T_b. (b) The distribution of the random vector (T_w, T_b). Are T_w and T_b independent? Consider both cases: drawing without and with replacement.

E 3.14 Random variables ξ and η are independent and have common uniform distribution on the interval $(c - \frac{1}{2}, c + \frac{1}{2})$. Find the density of the difference $\xi - \eta$.

E 3.15 Given independent random variables ξ and η with common distribution Q find $P(\xi = \eta)$. Consider both cases, where Q is first discrete and then continuous.

E 3.16 Is it true that the random vector (ξ, η) has a continuous distribution when both ξ and η do?

E 3.17 Find the distribution of the sum of independent random variables ξ, η if ξ has the uniform distribution on the interval $[0, 1]$ and $P(\eta = 0) = P(\eta = 1) = \frac{1}{2}$.

4 Parameters of Probability Distributions

In most practical situations we are not able to fully determine a probability distribution of a random variable or vector under consideration. Nevertheless, we can often obtain partial information on the distribution that suffices for our purpose. The most basic information about the random variable is given by the mean or mathematical expectation and the variance or its square root, which is known as the standard deviation. Other parameters such as higher moments are also considered, but are not as significant so they will be not covered in this book. The substantial role of the mean and variation partially explains the Law of Large Numbers which states convergence of the averages of random variables to their common mean. We will thus discuss these laws as well as various types of convergence of random variables. We will complete this Chapter with a problem on the correlation of two or more random variables.

4.1 Mathematical Expectation

The mathematical expectation, or simply expectation or mean, is the most basic parameter of most random variables. We often encounter such mathematical expectations in "real life", though we are rarely aware of it. We begin with an example.

Example 4.1.1 Mr. Smith invites you to the following game. You will pay him $4 for a single toss of his die. He will pay you $x for x pips on the die if $x > 1$. If x = 1 you will get one free extra toss and receive $x for x pips (even if x = 1 this time).

Who would win this game?

We can immediately see that there is no correct answer and the question should be changed to the following: for whom is the game profitable. The answer is still not obvious, at least for only a few games. For example, in a sequence of five tosses we may get:

$$\text{``5'', ``4'', ``5'', ``1'', extra toss ``6'', ``6''.}$$

Then, Mr. Smith wins $5 \cdot \$4 = \$20$, but pays $26. Still, the following sequence of pips may occur as well:

$$\text{``2'', ``2'', ``4'', ``3'', ``6'',}$$

which means a profit of $20 -$17 = $3 for Mr. Smith.

The whole situation changes when the game is repeated many times. Again, we are not able to say for sure who will win, but we can predict for whom the game should be profitable. Assume that Mr. Smith can find candidates for, say, 360 trials. Now, he may anticipate that any number will appear about 60 times and thus can perform the following forecast. First of all, he will receive $1440 = 360 \cdot $4 from the players. He will also pay for about sixty "2"s, sixty "3"s, ..., sixty "6"s, and for about sixty extra tosses will pay for some ten "2"s, ten "3"s, ..., ten "6"s. Mr. Smith will then pay approximately

$$10 \cdot 1 + 70 \cdot 2 + \ldots + 70 \cdot 6 = 10 + 70 \cdot 20 = 1410 \text{ (dollars)},$$

so he has a good chance to earn about $30.

One can ask what would happen for some other number of trials different from 360. In any case one can repeat the above calculation, but a more effective way is to compute the number m:

$$m = \frac{1}{36} \cdot 1 + \frac{7}{36} \cdot 2 + \frac{7}{36} \cdot 3 + \ldots + \frac{7}{36} \cdot 6 = \frac{47}{12} = 3.916(6),$$

and then to multiply it by the assumed number of trials. For example, we have $360m = 1410$, which is not surprising since the latter equality has been obtained form the former by dividing both sides by 360.

MAPLE ASSISTANCE 25

We can simulate the above game. Consider the following procedure game:

```
>   game := proc(n)
    global s, alist, average;
    local i, toss, die:
    alist := NULL:
    s := 0:
    for i from 1 to n do
    toss := rand(1..6):
    die := toss():
    if die = 1 then
    toss := rand(1..6):
    die := toss()
    fi:
    alist := alist,die:
    s := s + die:
    od:
    average := s/n:
    end:
```

Play the game twenty times:

```
>  game(20):
   alist;
   evalf(average);
```
$$4, 3, 4, 6, 5, 3, 6, 3, 2, 2, 2, 4, 4, 3, 3, 2, 4, 4, 6, 1$$
$$3.550000000$$

In this case Mr. Smith will have to pay \$3.55 on average for a single trial, so he will earn on average 45 cents each time. Now play the game 3,000 times (avoiding the details).

```
>  game(3000):
   evalf(average);
```
$$3.892000000$$

By the way, once we have the data we could use the stats package to compute the mean:

```
>  with(stats);
```
[*anova, describe, fit, importdata, random, statevalf, statplots, transform*]

```
>  evalf(describe[mean]([alist]));
```
$$3.892000000$$

We could help Mr. Smith in making his forecast.

```
>  xi := [seq(Weight(i,7/36),i=2..6),Weight(1,1/36)];
```

$$\xi := [\text{Weight}(2, \frac{7}{36}), \text{Weight}(3, \frac{7}{36}), \text{Weight}(4, \frac{7}{36}), \text{Weight}(5, \frac{7}{36}),$$
$$\text{Weight}(6, \frac{7}{36}), \text{Weight}(1, \frac{1}{36})]$$

```
>  describe[mean](xi);
```
$$\frac{47}{12}$$

```
>  evalf(%);
```
$$3.916666667$$

$$\cdots$$

In the above example the payoff after a single game is a random variable, say ξ. It takes one of six possible values:

$$x_1 = 1, \quad x_2 = 2, \quad \ldots, \quad x_6 = 6,$$

with probabilities

$$p_1 = \frac{1}{36}, \quad p_2 = \frac{7}{36}, \quad \ldots, \quad p_6 = \frac{7}{36}.$$

Now, the average payoff equals:

$$m = p_1 x_1 + p_2 x_2 + p_3 x_3 + p_4 x_4 + p_5 x_5 + p_6 x_6. \tag{4.1}$$

This quantity is called the mathematical expectation, the mean or the expected value of the random variable ξ. Mathematical expectation is just the sum of all possible values of ξ multiplied by their weights. This rule might be used as a general definition of mathematical expectation of random variables taking finite or at most countably many values. As we already have seen, other types of random variables are also considered. Fortunately, formula (4.1) can be generalized in terms of an integral with respect to measure (see Chapter 2).

Definition 4.1.1 *Let (Ω, Σ, P) be a probability space and let $\xi : \Omega \longrightarrow \mathbb{R}$ be a random variable. The quantity*

$$m = \mathbb{E}(\xi) = \mathbb{E}\xi = \int_\Omega \xi \, dP$$

is called the mathematical expectation, the mean or the expected value of the random variable ξ, provided the integral $\int_\Omega \xi \, dP$ exists (finite or infinite).

Example 4.1.2 Let random variable $\xi : \Omega \longrightarrow \mathbb{R}$ take finitely many values, say $x_1, \ldots, x_n$ with probabilities $p_1, \ldots, p_n$. Then the mean is

$$\mathbb{E}(\xi) = p_1 x_1 + \ldots + p_n x_n,$$

which is the same as (4.1). In fact, in this case ξ is a step function, namely $\xi = x_1 I_{A_1} + \ldots + x_n I_{A_n}$, where $A_i = \{\omega \in \Omega : \xi(\omega) = x_i\}$, for $i = 1, \ldots, n$. By assumption, $P(A_i) = p_i$, so by the definition of integral we obtain the above formula.

The theory of Lebesgue integration provides a list of useful properties of mathematical expectations.

Theorem 4.1.1 *Let ξ and η be random varibles defined on the probability space (Ω, Σ, P). Assume that the mathematical expectations $\mathbb{E}(\xi)$ and $\mathbb{E}(\eta)$ exist. Then:*

1. *If $\xi = const = c$, then $\mathbb{E}(\xi) = c$.*
2. *If $\xi \geq 0$, then $\mathbb{E}(\xi) \geq 0$.*
3. *For any real number α, $\mathbb{E}(\alpha\xi) = \alpha\mathbb{E}(\xi)$.*
4. *$\mathbb{E}(\xi + \eta) = \mathbb{E}(\xi) + \mathbb{E}(\eta)$.*
5. *$\mathbb{E}(\xi\cdot\eta)^2 \leq \mathbb{E}(\xi^2)\cdot\mathbb{E}(\eta^2)$, with the equality if and only if ξ and η are linearly dependent, i.e. there exist positive constants α, β such that $P(\alpha\xi + \beta\eta \neq 0) = 0$.*

As a corollary we have the following important theorem.

Theorem 4.1.2 *Let $\xi_1, \xi_2, \ldots \xi_n$ be random variables with a common mathematical expectation $m = \mathbb{E}(\xi_i)$ and define*

$$S_n = \xi_1 + \xi_2 + \ldots + \xi_n.$$

Then:

1. $\quad \mathbb{E}(S_n) = nm,$

2. $\quad \mathbb{E}\left(\dfrac{S_n}{n}\right) = m.$

4.2 Variance

Imagine that we have the following two sets of data representing the occurrence of two random variables which both have the mean $m = 16.8$.

MAPLE ASSISTANCE 26

```
>  data1 := [18.80, 18.80, 19.20,
   21.80, 17.40, 14.20, 14.60, 15.80, 14.40, 13.00]:
>  describe[mean](data1);
                16.80000000
>  data2 := [23.60, 23.40, 8.20,
   12.80, 14.20, 10.20, 13.20, 25.80, 13.40, 23.20]:
>  describe[mean](data2);
                16.80000000
```

Simple inspection shows that *data*1 is much more concentrated about the mean than *data*2. A good measure of concentration (or, equivalently, of scattering) about the mean is variance or its square root, which is known as standard deviation. In our example:

MAPLE ASSISTANCE 27

```
>  describe[variance](data1),describe[variance](data2);
                7.192000000, 37.65600000
>  describe[standarddeviation](data1),
   describe[standarddeviation](data2);
                2.681790447, 6.136448484
```

The formal definition is as follows.

Definition 4.2.1 *Let (Ω, Σ, P) be a probability space and let $\xi : \Omega \longrightarrow \mathbb{R}$ be a random variable. Assume that the mathematical expectation $m = \mathbb{E}(\xi)$ exists and is finite.*

The variance is defined as the mean of the square of the difference $\xi - m$ i. e.

$$\mathrm{Var}\,(\xi) = \mathbb{E}((\xi - m)^2) = \int_\Omega (\xi - m)^2 \, dP. \qquad (4.2)$$

The standard deviation is defined as

$$\sigma = \sigma(\xi) = \sqrt{\mathrm{Var}\,(\xi)}. \qquad (4.3)$$

One can ask why we have chosen the variance or standard deviation here instead of the more natural, at first glance, $\mathbb{E}(|\xi - m|)$, which is known as the *mean deviation*, to measure the concentration about the mean. The reason is that variance has nice analytical properties to be explained in the sequel. Moreover, we will see that it is particularly convenient to use standard deviation when we consider the normal (Gaussian) distribution, the most important of probability distributions.

4.2.1 Moments

The mean and variance are examples of moments and central moments. The moment of order $p = 1, 3, 4, \ldots$ is defined as

$$\mathbb{E}(\xi^p) = \int_\Omega \xi^p \, dP$$

and the central moment of order p as

$$\mathbb{E}((\xi - m)^p) = \int_\Omega (\xi - m)^p \, dP.$$

We derive another formula for the variance using some properties of the mean. Namely we have

$$\begin{aligned}
\mathrm{Var}\,(\xi) &= \mathbb{E}((\xi - m)^2) \\
&= \mathbb{E}(\xi^2 - 2m\xi + m^2) \\
&= \mathbb{E}(\xi^2) - 2m\mathbb{E}(\xi) + \mathbb{E}(m^2) \\
&= \mathbb{E}(\xi^2) - m^2.
\end{aligned}$$

and hence

$$\mathrm{Var}\,(\xi) = \mathbb{E}(\xi^2) - m^2, \tag{4.4}$$

which is particularly useful for calculations.

Also one can easily see that for any $c \in \mathbb{R}$

$$\mathrm{Var}\,(c\xi) = c^2 \mathrm{Var}\,(\xi). \tag{4.5}$$

The mathematical expectations and variances in the case of independent random variables have two very convenient properties.

Theorem 4.2.1 *Let ξ and η be independent random variables defined on a probability space (Ω, Σ, P).*

1. *If $\xi \geq 0$ and $\eta \geq 0$ or if $\mathbb{E}(|\xi|) < \infty$ and $\mathbb{E}(|\eta|) < \infty$, then the mathematical expectation of the product $\xi \cdot \eta$ exists and satisfies*

$$\mathbb{E}(\xi \cdot \eta) = \mathbb{E}(\xi) \cdot \mathbb{E}(\eta). \tag{4.6}$$

2. *If* $\mathrm{Var}\,(\xi) < \infty$ *and* $\mathrm{Var}\,(\eta) < \infty$, *then*

$$\mathrm{Var}\,(\xi + \eta) = \mathrm{Var}\,(\xi) + \mathrm{Var}\,(\eta). \tag{4.7}$$

Proof. Property 1. We present a proof in the special case when both random variables have discrete distributions concentrated at finitely many points. The general case can be proved by applying Fubini's Theorem, see Theorem 2.3.4. Assume that

$$P(\xi = x_i) = p_i, \qquad P(\eta = y_j) = q_j.$$

Then, the distribution of the product $\xi \cdot \eta$ is given as

$$P(\xi \cdot \eta = x_i y_j) = r_{ij},$$

and by independence we have

$$P(\xi \cdot \eta = x_i y_j) = P(\xi = x_i, \eta = y_j) = P(\xi = x_i) \cdot P(\eta = y_j) = p_i q_j.$$

Then:

$$\mathbb{E}(\xi \cdot \eta) = \sum_{i,j} r_{ij} x_i y_j = \sum_{i,j} p_i q_j x_i y_j = \sum_i p_i x_i \sum_j q_j y_j = \mathbb{E}(\xi)\mathbb{E}(\eta).$$

Property 2. Independence of ξ and η implies the independence of $\xi - \mathbb{E}(\xi)$ and $\eta - \mathbb{E}(\eta)$, see Theorem 3.4.1. Thus we have

$$\begin{aligned}
\mathrm{Var}\,(\xi + \eta) &= \mathbb{E}[((\xi + \eta) - \mathbb{E}(\xi + \eta))^2] \\
&= \mathbb{E}[((\xi - \mathbb{E}(\xi)) + (\eta - \mathbb{E}(\eta)))^2] \\
&= \mathrm{Var}\,(\xi) + \mathrm{Var}\,(\eta) + 2\mathbb{E}(\xi - \mathbb{E}(\xi))\,\mathbb{E}(\eta - \mathbb{E}(\eta)) \\
&= \mathrm{Var}\,(\xi) + \mathrm{Var}\,(\eta).
\end{aligned}$$

$\square$

We immediately see that the above theorem may be stated for any finite collection of independent random variables. Below, we present a special, yet important case.

Theorem 4.2.2 *Let* $\xi_1, \xi_2, \dots \xi_n$ *be independent random variables having the same standard deviation* σ. *Denote as before:*

$$S_n = \xi_1 + \xi_2 + \dots + \xi_n.$$

Then:

1. $\quad \mathrm{Var}\,(S_n) = n\sigma^2, \qquad \sigma(\xi) = \sigma\sqrt{n}.$

2. $\quad \mathrm{Var}\left(\dfrac{S_n}{n}\right) = \dfrac{\sigma^2}{n}, \qquad \sigma\left(\dfrac{S_n}{n}\right) = \dfrac{\sigma}{\sqrt{n}}.$

In other words, the standard deviation of the sum of n *independent random variables increases proportionally to* $\sqrt{n}$, *while standard deviation of the average of such components decreases like* $\dfrac{1}{\sqrt{n}}$.

4.3 Computation of Moments

The mean, the variance and other moments were defined in rather abstract way, so now we derive formulas that are more convenient for computing them. In particular, for discrete and continuous distributions these formulas will be easy to use. First, we state a fundamental theorem which says that the moments are uniquely determined by the probability distribution of the random variable or random vector.

Theorem 4.3.1 *Let (Ω, Σ, P) be a probability space, $X : \Omega \longrightarrow \mathbb{R}^n$ a random vector, and $g : \mathbb{R}^n \longrightarrow \mathbb{R}$ a Borel function. Let P_X denote the n-dimensional probability distribution of X. Recall that $g(X)$ denotes the composition $g \circ X$. Then*

$$\mathbb{E}(g(X)) = \int_{\mathbb{R}^n} g \, dP_X, \tag{4.8}$$

in the sense that if either of the two sides exists, then so does the other.

Proof. We only indicate the idea of the proof.
Assume first that g is an indicator i.e. $g = I_B$, $B \in \mathcal{B}(\mathbb{R})$. Then we have

$$\mathbb{E}(g(X)) = \int_{\Omega} I_B \circ X \, dP = \int_{\Omega} I_{X^{-1}(B)} \, dP = P(X^{-1}(B))$$

$$= P_X(B) = \int_B 1 \, dP_X = \int_{\mathbb{R}^n} g \, dP_X. \tag{4.9}$$

Now let g be a Borel step function, $g = \sum_{i=1}^n c_i I_{B_i}$. By the linearity property of integrals, formula (4.8) holds true also for this function.

The final two steps extend formula (4.8) to a nonnegative Borel function and then through a limiting process to any Borel functions in the definition of integral, see Definition 2.2.2. $\qquad\square$

Using the above theorem one can derive formulas for the mean and the variance. Namely, putting $X = \xi$ and $g(x) = x$ for $x \in \mathbb{R}$ we have

$$\mathbb{E}(\xi) = \int_{\mathbb{R}} x \, dP_\xi(x), \tag{4.10}$$

for any random variable ξ. Putting $g(x) = x^k$ we get a formula for the moment of order k:

$$\mathbb{E}(\xi^k) = \int_{\mathbb{R}} x^k \, dP_\xi(x). \tag{4.11}$$

Similarly,

$$\mathrm{Var}\,(\xi) = \int_{\mathbb{R}} (x - m)^2 \, dP_\xi(x) = \int_{\mathbb{R}} x^2 \, dP_\xi - m^2, \tag{4.12}$$

where $m = \mathbb{E}(\xi)$.

. . .

In the discrete and continuous cases Theorem 4.3.1 implies the following more practical formulas.

Theorem 4.3.2 *Let (Ω, Σ, P) be a probability space, $X : \Omega \longrightarrow \mathbb{R}^n$ a random vector having discrete distribution determined by points $\{x_i\}$ and probabilities $\{p_i\}$, $i = 0, \ldots, N$, where $N \leq \infty$. Let $g : \mathbb{R}^n \longrightarrow \mathbb{R}$ be a Borel function. Then*

$$\mathbb{E}(g(X)) = \sum_{i=0}^{N} g(x_i)p_i, \qquad (4.13)$$

in the sense that if either of the two sides exists, then the other does too. For the case $N = \infty$ the infinite sum is interpreted here in a similar sense to the definition of integral, i.e. $\sum_{n=0}^{\infty} a_n := \sum_{n=0}^{\infty} a_n^+ - \sum_{n=0}^{\infty} a_n^-$, where $a_n = a_n^+ - a_n^-$ and $a_n^+, a_n^- \geq 0$.

Theorem 4.3.3 *Let (Ω, Σ, P) be a probability space and let $X : \Omega \longrightarrow \mathbb{R}^n$ be a random vector having a continuous distribution with density f. Let $g : \mathbb{R}^n \longrightarrow \mathbb{R}$ be a Borel function. Then*

$$\mathbb{E}(g(X)) = \int_{\mathbb{R}^n} g(x)f(x)\,dx, \qquad (4.14)$$

in the sense that if either of the two sides exists, then so does the other, while dx means integration with respect to the Lebesgue measure.

These theorems are proved in a similar manner to Theorem 4.3.1.

As before we obtain formulas for the mean and the variance:
discrete case:

$$\mathbb{E}(\xi) = \sum_{i=0}^{N} x_i p_i, \qquad \text{Var}\,(\xi) = \sum_{i=0}^{N}(x_i - m)^2 p_i, \qquad (4.15)$$

continuous case:

$$\mathbb{E}(\xi) = \int_{\mathbb{R}} x f(x)\,dx, \qquad \text{Var}\,(\xi) = \int_{\mathbb{R}}(x - m)^2 f(x)\,dx. \qquad (4.16)$$

Convince yourself that formula (4.15) is compatible with Example 4.1.2.

4.3.1 Evaluation of the Mean and Variance

The above formulas will be used to calculate the mean and the variance of a number of distributions for the random variables in Chapter 5. At this point we show only some simple examples of applications of these formulas. In particular, we will also show how to compute the mean and the variance of certain random variables without knowing their distribution.

Example 4.3.1 A random variable ξ equal to a constant m has one point distribution concentrated at m. Then

$$\mathbb{E}(\xi) = m, \qquad\qquad \mathrm{Var}\,(\xi) = 0. \qquad\qquad (4.17)$$

Example 4.3.2 A random variable ξ taking exactly two values: 1 with probability p and 0 with probability $q = 1 - p$, i.e. ξ has the two point distribution with parameters p and q. Then

$$\mathbb{E}(\xi) = p, \qquad\qquad \mathrm{Var}\,(\xi) = pq. \qquad\qquad (4.18)$$

Example 4.3.3 For a random variable ξ having uniform distribution on the interval (a, b), i.e. $\xi \in U(a, b)$, we have

$$\mathbb{E}(\xi) = \frac{a+b}{2}, \qquad\qquad \mathrm{Var}\,(\xi) = \frac{(b-a)^2}{12}. \qquad\qquad (4.19)$$

Example 4.3.4 How long should one toss a die to get "6"?

Certainly, there is no unique answer to this question, see also Example 1.2.3. On the other hand one can determine the mean waiting time for first "6". Let N be the number of tosses in which "6" appears for the first time. We will find the distribution of the random variable N. We have to compute the probabilities $P(N = k)$, for $k = 1, 2, 3, \ldots$. Note that the event $\{N = k\}$ is just the intersection of k independent events such that "6" does not appear in any of the first $k - 1$ tosses (and has probability $5/6$) and then appears in the k-th toss (this has probability $1/6$). So we have

$$P(N = k) = \left(\frac{5}{6}\right)^{k-1} \frac{1}{6}.$$

Therefore

$$m := \mathbb{E}(N) = \sum_{k=1}^{\infty} \left(\frac{5}{6}\right)^{k-1} \frac{1}{6} k$$

and

$$\mathrm{Var}\,(N) = \sum_{k=1}^{\infty} \left(\frac{5}{6}\right)^{k-1} \frac{1}{6} k^2 - m^2.$$

We evaluate m and $\mathrm{Var}\,(N)$ using MAPLE.

MAPLE ASSISTANCE 28
```
> m := sum('(5/6)^(k-1)*k/6', 'k'=1..infinity);
> var := sum('(5/6)^(k-1)*k^2/6', 'k'=1..infinity) - m^2;
                        m := 6
                        var := 30
```

In the following example there is no easy way to determine the distribution of the random variable, but we can still compute the mean and the variance of this variable.

Example 4.3.5 Imagine we are drawing with replacement units from the population of size N. (We can meet different people at the party, collect tickets or choose by chance TV channels). We want to know $\mathbb{E}(T)$, the mean time needed to obtain exactly r distinct units. The random variable T takes infinitely many values and it is not obvious how to determine its distribution (though it can be done). On the other hand, let us note that T is a sum of simpler random variables. In fact, if we already have n distinct units, the waiting time T_n for the next distinct unit has its distribution very much like the distribution of the waiting time for the first "6" in the previous example. As before, we can see that

$$P(T_n = k) = \left(\frac{n}{N}\right)^{k-1} \frac{N-n}{N}, \quad k = 1, 2, 3, \ldots$$

This is all we need to compute the mean of all T_n. Then, using property 4 of Theorem 4.1.1, we have

$$\mathbb{E}(T) = \mathbb{E}(1 + T_1 + \ldots + T_{r-1}) = 1 + \mathbb{E}(T_1) + \ldots + \mathbb{E}(T_{r-1}).$$

We can also compute the variance of any T_n and then apply property 2 of Theorem 4.2.1 to obtain $\operatorname{Var}(T)$ since the waiting times $T_1, \ldots, T_{r-1}$ are independent random variables.

We ask MAPLE for help in performing the above computations:

MAPLE ASSISTANCE 29

First we calculate the mean and the variance of a random variable taking values 1, 2, 3, ... with probabilities p, $(1-p)p$, $(1-p)^2 p$, ...

```
>  m := sum('(1-p)^(k-1)*p*k', 'k'=1..infinity);
```

$$m := \frac{1}{p}$$

```
>  var := sum('(1-p)^(k-1)*p*k^2', 'k'=1..infinity) - m^2;
```

$$var := \frac{-p+2}{p^2} - \frac{1}{p^2}$$

```
>  simplify(%);
```

$$-\frac{-1+p}{p^2}$$

We are interested in the case $p = \dfrac{N-n}{N}$.

```
>  p := (N - n)/N
```

Compute $\mathbb{E}(T_n)$ and $\operatorname{Var}(T_n)$.

```
>  M := unapply(m,(n,N));
```

$$M := (n, N) \to \frac{N}{N-n}$$

```
> Var := unapply(var,(n,N));
```

$$Var := (n, N) \rightarrow \frac{N^2\left(-\dfrac{N-n}{N}+2\right)}{(N-n)^2} - \frac{N^2}{(N-n)^2}$$

Define $\mathbb{E}(T)$ and Var (T).

```
> ET := (N,r) -> sum(M(i,N),i=0..r-1);
```

$$ET := (N, r) \rightarrow \sum_{i=0}^{r-1} \mathrm{M}(i, N)$$

```
> V := (N,r) -> sum(Var(i,N), i=0..r-1);
```

$$V := (N, r) \rightarrow \sum_{i=0}^{r-1} \mathrm{Var}\,(i, N)$$

At last, find $\mathbb{E}(T)$ and Var (T) for specific values of N and r.

```
> evalf(ET(100,30)); evalf(V(100,30));
                      35.45407600
                      6.885850949
> evalf(ET(200,100)); evalf(V(200,100));
                      138.1306861
                      60.37514711
> evalf(ET(200,190)); evalf(V(200,190));
                      589.8125388
                      3017.340055
```

Let us look more carefully at the above results. If the number of distinct units r we want to obtain is relatively small compared to the population size, then the number of drawings is only slightly greater than r and the variance is moderate. If the number r is comparable to the population size we need many more drawings and the large variance indicates that the values of T may be far away from its mean.

The idea underlying this example can be applied if one wants to determine the size of the population by examining just a few of its elements. For example, if one finds 8 distinct units out of 12 drawn from a population, there is a good chance that the size of the population is not greater than 100.

MAPLE ASSISTANCE 30

```
> evalf(ET(100,8)); evalf(V(100,8));
                      8.294833858
                      .3105547438
```

As we will see soon, $P(T \geq 12) \leq 0.0226$ for $N = 100$ and $r = 8$. In other words, if the size of the population is 100 and we want to get 8 distinct elements, then the probability that we will need 12 or more drawings is 0.0226.

There are statistical methods allowing the estimation of the true population size. What is even more important, statistics provides a precise formulation of the problem.

$$\cdots$$

Example 4.3.6 Find the expected value and the variance of the area of the disk centered at $(0,0)$ with radius $r = \sqrt{\xi^2 + \eta^2}$, where ξ and η are randomly chosen from the interval $(0,1)$.

Let S denote the area. Clearly, $S = \pi(\xi^2 + \eta^2)$. According to (4.14) we have

$$\mathbb{E}(S) = \int_{\mathbb{R}^2} \pi(x^2 + y^2) f(x,y)\, d(x,y),$$

$$\mathbb{E}(S^2) = \int_{\mathbb{R}^2} \pi(x^2 + y^2)^2 f(x,y)\, d(x,y)$$

and then

$$\mathrm{Var}\,(S) = \mathbb{E}(S^2) - \mathbb{E}(S)^2.$$

Here f is the density of the random vector (ξ, η). Assuming that ξ and η are independent and have uniform distributions on $(0,1)$, we know that f is the product of the indicator functions of the unit interval and so is the indicator function of the unit square.

MAPLE ASSISTANCE 31

```
>  S := (x,y) -> Pi*(x^2 + y^2);
```
$$S := (x, y) \rightarrow \pi\,(x^2 + y^2)$$
```
>  int(S(x,y),y=0..1);
```
$$\frac{1}{3}\pi + \pi x^2$$
```
>  E(S) := int(%,x=0..1);
```
$$\mathbb{E}(S) := \frac{2}{3}\pi$$

We will be more concise in computing the variance.
```
>  Var(S) := int(int(S(x,y)^2,y=0..1),x=0..1) - E(S)^2;
```
$$\mathrm{Var}(S) := \frac{8}{45}\pi^2$$

We complete this Section with an example showing that the mean of a random variable may be infinite.

Example 4.3.7 A player tosses a coin until a head appears and then the player gains 2^k dollars, where k is the number of tosses he has made. We want to know his expected payoff. It would be really good to know the answer, because we could then invite the readers to participate in such a game, and then winning from them m dollars for each trial, where m is the mean we are looking for. Such a game would be fair.

Let W denote the payoff. It takes values $2, 4, 8, \ldots$ with probabilities $\frac{1}{2}, \frac{1}{4}, \frac{1}{8}, \ldots$. So

$$\mathbb{E}(W) = \sum_{k=1}^{\infty} 2^k \frac{1}{2^k} = \infty$$

and we have to give up our plans.

4.4 Chebyshev's Inequality

The meaning of central moments, especially that of the variance, is partly explained by the following theorem. Generally speaking, it says that if the central moments are small, then the random variable cannot deviate much from its mean.

Theorem 4.4.1 Let (Ω, Σ, P) be a probability space, $\xi : \Omega \longrightarrow \mathbb{R}$ a random variable, $\varepsilon > 0$ a fixed number.

1. Let $\xi \geq 0$ a.e. (see page 43) and let k be a natural number. Then

$$P(\xi \geq \varepsilon) \leq \frac{\mathbb{E}(\xi^k)}{\varepsilon^k}. \tag{4.20}$$

2. Let ξ be a random variable with finite mean m. Then

$$P(|\xi - m| \geq \varepsilon) \leq \frac{\operatorname{Var}(\xi)}{\varepsilon^2}. \tag{4.21}$$

3. Let ξ be a random variable with finite mean m and finite standard deviation $\sigma > 0$. For any positive $c > 0$ we have

$$P(|\xi - m| \geq c\sigma) \leq \frac{1}{c^2}. \tag{4.22}$$

Each of the above inequalities is known as Chebyshev's Inequality.

Proof. 1. Since $\xi \geq 0$ the basic properties of the integral imply

$$\mathbb{E}(\xi^k) = \int_{\Omega} \xi^k \, dP \geq \int_{\{\omega : \xi(\omega) \geq \varepsilon\}} \xi^k \, dP \geq \int_{\{\omega : \xi(\omega) \geq \varepsilon\}} \varepsilon^k \, dP = \varepsilon^k P(\xi \geq \varepsilon),$$

which proves (4.20).

2. It is enough to apply the inequality we have just proved to the random variable $|\xi - m|$ putting $k = 2$.

3. In the last inequality put $\varepsilon = c\sigma$ and apply property (4.5).   □

Let us have a closer look at inequality (4.22). It says, for example, that the values of any random variable with finite mean m and finite variance σ^2 belong to the interval $(m - 3\sigma, m + 3\sigma)$ with probability at least $\frac{8}{9}$, and to the interval $(m - 5\sigma, m + 5\sigma)$ with probability ≥ 0.96. The above estimates hold for any random variable no matter what probability distribution it has. If we know the distribution then the estimates may be much better. We refer the reader to page 140, where similar estimates are given for a normally distributed (Gaussian) random variable.

We now show some simple applications of Chebyshev's Inequality.

Example 4.4.1 Common sense indicates that in a long series of tosses of a "fair" coin we can expect a similar number of heads and tails. We shall justify this by estimating the probability that the difference between the two numbers is greater than some constant. For simplicity assume that 1000 tosses were made. Let ξ denote the number of heads, so $1000-\xi$ is the number of tails. Let us note that ξ is the sum, $\xi = \sum_{i=1}^{1000} \xi_i$, of independent random variables ξ_i having two point distributions, with

$$P(\xi_i = 0) = \frac{1}{2}, \quad P(\xi_i = 1) = \frac{1}{2}, \quad \text{for } i = 1, \ldots, 1000.$$

We can interpret each ξ_i as the appearance of a head at the i-th toss. Hence, $\mathbb{E}(\xi) = 1000 \cdot \frac{1}{2} = 500$ and $\text{Var}\,(\xi) = 1000 \cdot \frac{1}{2} \cdot \frac{1}{2} = 250$.

From inequality (4.21) we then have

$$P(|\xi - (1000 - \xi)| \geq 100) = P(|\xi - 500| \geq 50) \leq \frac{250}{50^2} = 0.1.$$

One may argue that this probability seems too large. Again, common sense says that it should be considerably less. In fact, we will show later in Example 5.4.3 using the Central Limit Theorem that $P(|\xi - 500| \geq 50) \leq 0.0016$. In using the Chebyshev's Inequality we have not used any information on the distribution of ξ, so our result here is a much rougher estimate.

Example 4.4.2 *Example 4.3.5 continued*

Let $N = 100$, $r = 8$. Estimate the probability that $T \geq 12$.

$T \geq 12$ is equivalent to $T - \mathbb{E}(T) \geq 12 - \mathbb{E}(T)$ and this implies $|T - \mathbb{E}(T)| \geq 12 - \mathbb{E}(T)$. Hence, from Chebyshev's Inequality we have

$$P(T \geq 12) = P(T - \mathbb{E}(T) \geq 12 - \mathbb{E}(T)) \leq P(|T - \mathbb{E}(T)| \geq 12 - \mathbb{E}(T))$$

$$\leq \frac{\text{Var}\,(T)}{(12 - \mathbb{E}(T))^2}.$$

MAPLE ASSISTANCE 32

```
>  V(100,8)/(12 - ET(100,8))^2;
                  33724884017383717847705
                  ───────────────────────
                  149082844565335473824329
>  evalf(%,3);
                        .0226
```

So the probability is estimated by 0.0226.

Example 4.4.3 *Example 4.3.5 continued*
How many drawings with replacement are needed to obtain 100 distinct units from a population of size 200 with at least probability 0.95?
We have to find x, a number of drawings, such that $P(T < x) > 0.95$. We may certainly assume that $x > m = \mathbb{E}(T)$. Let $\varepsilon = x - m$. We have

$$P(T < x) = P(T < m + \varepsilon) = 1 - P(T \geq m + \varepsilon)$$
$$\geq 1 - P(|T - m| \geq \varepsilon) \geq 1 - \frac{\text{Var}\,(T)}{\varepsilon^2}.$$

Now, it is enough to choose x such that $1 - \dfrac{\text{Var}\,(T)}{(x - m)^2} \geq 0.95$.

MAPLE ASSISTANCE 33

```
>  m := evalf(ET(200,100)): var := evalf(V(200,100)):
>  solve(1 - var/(x-m)^2 >0.95,x);
```
RealRange($-\infty$, Open(103.3815431)), RealRange(Open(172.8798291), ∞)

So 173 drawings are enough to have a certainty of 95% that we will get 100 distinct units. Allowing a certainty of 90% we can compute that 163 drawings suffice. Actually, one can improve on these results (see page 154).

4.5 Law of Large Numbers

Before the 1930s when the modern definition of probability become firmly established, this notion had been understood as follows. If one wants to determine the probability p of some event, say A, one should repeat many times, say N times, the experiment involving this event A, counting the experiments that result in A. Dividing the number of successes obtained, S_N by N provides us with an approximation of the probability of A. More precisely, probability was understood as the limit

$$p = P(A) = \lim_{N \to \infty} \frac{S_N}{N}. \tag{4.23}$$

The modern, axiomatic, definition of probability based on the measure theory, and used in this book, might be seem to be less natural. Still, it has proven to be very useful during the last 70 years. In particular, the axiomatic definition based on the theory of measure allowed powerful analytical tools to be used in the calculus of probability and in statistics. Moreover, the Law of Large Numbers will say that the modern definition extends in certain way the old one, see Theorem 4.5.3 below for an appropriate result and more explanation. The old definition is now known as the frequency definition of probability.

To have more insight into the Law of Large Numbers note that in some previous MAPLE examples we repeated some experiments many times to discover a particular pattern or rule. Behind such a procedure is our belief that as the number of experiments increases the corresponding pattern will become more and more apparent. For example, in tossing a die 10 times we are not surprised if we do not obtain a "6". On the other hand, having no "6" after 100 tosses seems to be almost impossible.

MAPLE ASSISTANCE 34
```
> evalf((5/6)^10), evalf((5/6)^100);
                .1615055829, .1207467347 10^-7
```
In fact, we believe that in 100 tosses one should obtain at least, say 10 "6"s, and actually this number should oscillate about $\dfrac{100}{6}$. Such reasoning is justified by the Law of Large Numbers. They say in fact that if the number of tosses N is large, then the number of "6"s will oscillate about $\dfrac{N}{6}$. We will establish and prove the so called Weak Law of Large Numbers and then indicate what the Strong Law of Large Numbers is.

Since the Laws of Large Numbers are expressed in terms of the convergence of arithmetic averages of sequences of random variables we will state and compare definitions of convergence of random variables.

4.5.1 The Weak Law of Large Numbers

In the sequel we will use the following notation for the sum of random variables.

$$S_n = \xi_1 + \ldots + \xi_n.$$

Theorem 4.5.1 Let (Ω, Σ, P) be a probability space and let $\xi_1, \xi_2, \xi_3, \ldots$ be a sequence of independent random variables defined on this space having finite means and variances with a common bound, i.e. there is an $M \in \mathbb{R}$ such that $\mathrm{Var}\,(\xi_i) \leq M$ for all i. Then, for every $\varepsilon > 0$,

$$\lim_{n \to \infty} P\left(\left|\frac{S_n - \mathbb{E}(S_n)}{n}\right| \geq \varepsilon\right) = 0.$$

Proof. The independence of the random variables yields

$$\text{Var}\,(S_n) = \text{Var}\,(\xi_1) + \ldots + \text{Var}\,(\xi_n) \le nM.$$

Chebyshev's Inequality (4.21) applied to the variable S_n gives

$$P\left(\frac{|S_n - \mathbb{E}(S_n)|}{n} \ge \varepsilon\right) = P(|S_n - \mathbb{E}(S_n)| \ge n\varepsilon) \le \frac{\text{Var}\,(S_n)}{(n\varepsilon)^2} \le \frac{M}{n\varepsilon^2},$$

which proves the theorem.   □

The following theorems are immediate consequences of the above.

Theorem 4.5.2 *Assume additionally that the means are equal to m. Then, for every $\varepsilon > 0$,*

$$\lim_{n\to\infty} P\left(\left|\frac{S_n}{n} - m\right| \ge \varepsilon\right) = 0.$$

This theorem is sometimes meant as follows: $\xi_n(\omega)$ is the numerical value of some physical quantity ξ measured at a point ω in time interval n. Then $\dfrac{S_n}{n}$ is the time average mean, while $m = \mathbb{E}(\xi)$ is the space average or space mean. Theorem 4.5.2 says that asymptotically these two means are the same.

Theorem 4.5.3 *Let S_n denote the number of successes in a sequence of independent Bernoulli trials, each with the same probability of success p. Then, for every $\varepsilon > 0$,*

$$\lim_{n\to\infty} P\left(\left|\frac{S_n}{n} - p\right| \ge \varepsilon\right) = 0.$$

The last formula just says that the frequency definition of probability is compatible with the measure theoretic approach. In fact, consider an event $A \in \Sigma$ and consider occurrence of A as a success. Then, formula (4.5.3) states that the probability $p = P(A)$ of success is asymptotically equal to $\frac{S_n}{n}$ in a certain sense.

4.5.2 Convergence of Random Variables

The Weak Law of Large Numbers actually says that a certain sequence of random variables is convergent. Still, it is not pointwise convergence as we have said nothing on convergence of specific realisation values, but it is rather a convergence of probabilities. The following definition makes it precise.

Definition 4.5.1 *Let (Ω, Σ, P) be a probability space and $\xi_1, \xi_2, \xi_3, \ldots$ be a sequence of random variables defined on this space.*

We say that ξ_n converges stochastically (or in probability) to the random variable $\xi : \Omega \longrightarrow \mathbb{R}$, and denote it by

$$\xi_n \xrightarrow{\,p\,} \xi,$$

if for every $\varepsilon > 0$

$$\lim_{n \to \infty} P(|\xi_n - \xi| \geq \varepsilon) = 0.$$

Now, the Law of Large Numbers may be reformulated as follows. The statements of Theorem 4.5.1, Theorem 4.5.2 and Theorem 4.5.3 are just

$$\frac{S_n - \mathbb{E}(S_n)}{n} \xrightarrow{\,p\,} 0, \quad \frac{S_n}{n} \xrightarrow{\,p\,} m, \quad \text{and} \quad \frac{S_n}{n} \xrightarrow{\,p\,} p.$$

Another type of convergence is convergence with probability one.

Definition 4.5.2 *Let (Ω, Σ, P) be a probability space and $\xi_1, \xi_2, \xi_3, \ldots$ be a sequence of independent random variables defined on this space.*

We say that ξ_n converges with probability 1 (or almost sure) to the random variable $\xi : \Omega \longrightarrow \mathbb{R}$, and denote

$$\xi_n \xrightarrow{\,1\,} \xi,$$

if

$$P(\{\omega : \lim_{n \to \infty} \xi_n(\omega) = \xi(\omega)\}) = 1.$$

Let us compare these two kinds of convergence. First we state without proof the following:

Theorem 4.5.4 *Let (Ω, Σ, P) be a probability space, $\xi : \Omega \longrightarrow \mathbb{R}$ a random variable, and $\xi_1, \xi_2, \xi_3, \ldots$ a sequence of random variables defined on Ω.*
If $\xi_n \xrightarrow{\,1\,} \xi$, then $\xi_n \xrightarrow{\,p\,} \xi$.

On the other hand, the converse is not true, as the following example shows.

Example 4.5.1 Let: $\Omega = [0, 1]$, $\Sigma = \mathcal{B}(0, 1)$ and let P be the Lebesgue measure. Consider the sequence of random variables $\xi_{11}, \xi_{21}, \xi_{22}, \xi_{31}, \ldots$, defined on Ω as follows:

$$\xi_{kl}(\omega) = \begin{cases} 1, & \text{for } \frac{l-1}{k} < \omega \leq \frac{l}{k} \\ 0, & \text{otherwise}, \end{cases}$$

where $k = 1, 2, 3, \ldots$, $l = 1, \ldots k$. For any $0 < \varepsilon < 1$ we have $P(|\xi_{kl}| \geq \varepsilon) = \frac{1}{k}$. Hence our sequence is stochastically convergent to zero in the probability space (Ω, Σ, P). On the other hand, for any fixed ω the sequence $\xi_{kl}(\omega)$ contains infinitely many 0's and infinitely many 1's, so it is not convergent. It clearly implies that ξ_{kl} cannot be convergent with probability 1 to any limit.

MAPLE ASSISTANCE 35   We will animate the sequence ξ_{kl} above for $k = 4$ and $k = 9$.

```
> k := 4:
> for l from 1 to k do
> xi[l] := x->piecewise((l-1)/k < x and x <= l/k,1)
  od:
> rys.($1..k) := seq(plots[display](plot(xi[l],0..1,
  discont = true,color=BLACK,xtickmarks=[0.5,1],
  ytickmarks=[0.5,1],thickness=2)),l =1..k):
  plots[display](rys.($1..k),insequence=true);
```

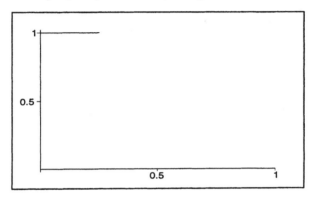

MAPLE has built up a sequence of four plots (first plot is shown only) that may be animated on the computer screen with the help of the mouse. For our purpose, we are displaying the whole sequence.

```
> plots[display](%);
```

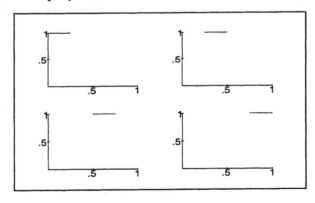

Let $k = 9$. Now, we will skip one plot.

```
> k := 9:
> for l from 1 to k do
```

```
>  xi[1] := x->piecewise((1-1)/k < x and x <= 1/k,1) od:
>  rys.($1..k) := seq(plots[display](plot(xi[1],0..1,
   discont = true,color=BLACK,xtickmarks=[0.5,1],
   ytickmarks=[0.5,1],thickness=2)),1 =1..k):
   plots[display](rys.($1..k),insequence=true):
>  plots[display](%);
```

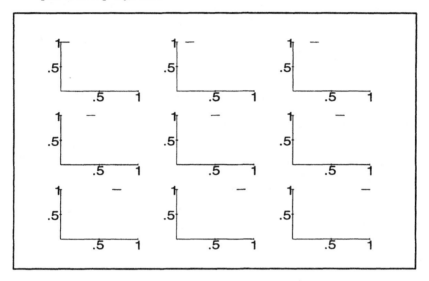

Other types of convergence of a sequence of random variables are also considered. They are important especially in the theory of stochastic processes.

Definition 4.5.3 *Let ξ_1, ξ_2, ξ_3, ... be a sequence of random variables having distribution functions F_1, F_2, F_3, ...*

We say that ξ_n converges in distribution to the random variable ξ, and denote it by

$$\xi_n \xrightarrow{d} \xi,$$

if

$$\lim_{n \to \infty} F_n(a) = F_\xi(a),$$

for any point $a \in \mathbb{R}$ such that F is continuous in a.

Let $p > 0$ be fixed.

Definition 4.5.4 *Let (Ω, Σ, P) be a probability space and $\xi_1, \xi_2, \xi_3, ...$ be a sequence of random variables defined on this space.*

We say that ξ_n converges in L^p to the random variable $\xi : \Omega \to \mathbb{R}$, and denote it by

$$\xi_n \xrightarrow{L^p} \xi,$$

if

$$\lim_{n \to \infty} \mathbb{E}(|\xi_n - \xi|^p) = 0$$

In particular, if $p = 1$, then we talk about convergence in mean, and if $p = 2$, we talk about mean-square convergence.

Even more types of convergence is considered and there exist relationships between most of them like this stated in Theorem 4.5.4. We will indicate to a few of them in Section 7.5. For now, we just mention that convergence in probability implies convergence in distribution.

4.5.3 The Strong Law of Large Numbers

As we said before, the estimation given by Chebyshev's Inequality was of moderate use, so it is not a surprise that the Weak Law of Large Numbers, proved by using this inequality, may be remarkably improved. In fact, using the Kolmogorov Inequality and some other analytical tools one can establish the so called Strong Law of Large Numbers, which gives conditions for convergence with probability 1 of the arithmetic averages of independent random variables. We state without proofs three versions of such theorems.

Theorem 4.5.5 *Let (Ω, Σ, P) be a probability space and let $\xi_1, \xi_2, \xi_3, \dots$ be a sequence of independent random variables defined on Ω with finite means and variances satisfying the condition:*

$$\sum_{i=1}^{\infty} \frac{\mathrm{Var}\,(\xi_i)}{i^2} < \infty.$$

Then

$$\frac{S_n - \mathbb{E}(S_n)}{n} \xrightarrow{1} 0.$$

Theorem 4.5.6 *Assume additionally that all the means are equal to m. Then*

$$\frac{S_n}{n} \xrightarrow{1} m.$$

If the random variables ξ_i have a common probability distribution we do not need any further assumption on the variance.

In the sequel we will use abbreviation *i.i.d.* for a sequence of independent random variables that are identically distributed.

Theorem 4.5.7 *Let (Ω, Σ, P) be a probability space and let $\xi_1, \xi_2, \xi_3, \dots$ be an i.i.d. sequence. Assume that the common mean m is finite. Then*

$$\frac{S_n}{n} \xrightarrow{1} m.$$

Example 4.5.2 *Example 4.1.1 continued.*

We will simulate the game (1) 20 series of 200 tosses each, (2) 20 series of 1000 tosses each. For each series we determine the realisation of $\dfrac{S_n}{n} - m$, where n is 200 and 1000, respectively.

MAPLE ASSISTANCE 36

First check that the procedure **game** is still active. If it is not, we should run this procedure again.

```
>  res1 := NULL:
>  from 1 to 20 do
   game(200):
   res1 := res1,evalf(average - 47/12,3)
   od:
   res1;
```

.0483, −.0867, −.207, .123, .153, .163, .0283, −.0167, −.0967, −.132, .0383, −.0117, −.107, −.137, .0883, .103, −.237, −.227, −.167, .00333

```
>  res2 := NULL:
   from 1 to 20 do
   game(1000):
   res2 := res2,evalf(average - 47/12,3)
   od:
   res2;
```

.0683, −.00467, .00933, −.0317, −.0537, .0383, −.0317, −.0837, −.00867, −.0587, −.0687, .0493, .0313, −.0507, −.0377, .0113, −.103, −.0287, .0613, −.0617

We can see here that, generally, the differences in the first case were greater than in the second one. Also, we can see it in a more accurate way.

```
>  add(res1[i]^2,i=1..nops([res1])),
   add(res2[i]^2,i=1..nops([res2]));
                   .3314862089, .0529072667
>  max(op(map(abs,[res1]))),max(op(map(abs,[res2])));
                   .237, .103
```

4.6 Correlation

The two parameters considered above, the mean and the variance, contained information about a single random variable, or more precisely, about its distribution. There are also parameters characterising the dependence of two or more random variables. Two such parameters, the covariance and the coefficient of correlation are discussed below. The idea is that if two random variables are independent then their covariance is zero, while if random variables are strongly linearly related, then their covariance is as large as possible. The coefficient of correlation is just the covariance normalised to one.

4.6.1 Marginal Distributions

Let $\xi, \eta : \Omega \longrightarrow \mathbb{R}$ be random variables defined on the same probability space (Ω, Σ, P) and let $P_{(\xi, \eta)}$ denote the distribution of the random vector (ξ, η). We can determine the distributions of ξ and η, P_ξ and P_η, respectively, which in this context are called the marginal distributions. Namely, we have

$$P_\xi(A) = P_{(\xi, \eta)}(A \times \mathbb{R}), \qquad P_\eta(B) = P_{(\xi, \eta)}(\mathbb{R} \times B),$$

for any Borel sets $A, B \subset \mathbb{R}$.

In the most important cases when $P_{(\xi, \eta)}$ is discrete or continuous we can easily derive the following useful formulas for the marginal distributions.

Theorem 4.6.1 *We have:*

1. *Let $P_{(\xi, \eta)}$ have the discrete distribution $P_{(\xi, \eta)}(x_i, y_j) = r_{ij}$. Then the marginal distributions are also discrete and*

$$P_\xi(x_i) = \sum_j r_{ij} := p_i, \qquad P_\eta(y_j) = \sum_i r_{ij} := q_j \qquad (4.24)$$

2. *Let $P_{(\xi, \eta)}$ have a continuous distribution with density $h : \mathbb{R}^2 \longrightarrow \mathbb{R}$. Then the marginal distributions are also continuous and have densities*

$$f(x) = \int_\mathbb{R} h(x, y)\, dy, \qquad g(y) = \int_\mathbb{R} h(x, y)\, dx. \qquad (4.25)$$

4.6.2 Covariance and the Coefficient of Correlation

Let $\xi, \eta : \Omega \longrightarrow \mathbb{R}$ be as above, and having finite means $m_\xi = \mathbb{E}(\xi)$, $m_\eta = \mathbb{E}(\eta)$ and variances $\sigma_\xi^2 = \mathrm{Var}\,(\xi)$, $\sigma_\eta^2 = \mathrm{Var}\,(\eta)$. Also assume that ξ and η are both not constant, which exactly means that $\sigma_\xi^2 > 0$, $\sigma_\eta^2 > 0$.

Definition 4.6.1 *The number*

$$cov(\xi, \eta) = \mathbb{E}((\xi - m_\xi)(\eta - m_\eta))$$

is called the covariance of ξ and η.

The basic properties of covariance are summarised below.

Theorem 4.6.2 *We have:*

1. *If ξ and η are independent, then $cov(\xi, \eta) = 0$.*
2. $\mathrm{Var}\,(\xi + \eta) = \mathrm{Var}\,(\xi) + \mathrm{Var}\,(\eta) + 2cov(\xi, \eta)$.
3. $|cov(\xi, \eta)| \leq \sigma_\xi \sigma_\eta$.
4. $|cov(\xi, \eta)| = \sigma_\xi \sigma_\eta$ *if and only if $\eta = a\xi + b$, where $a, b \in \mathbb{R}$, with $a \neq 0$. In that case, $sign\,(cov(\xi, \eta)) = sign\,(a)$.*
5. $cov(\xi, \eta) = \mathbb{E}(\xi\eta) - m_\xi m_\eta$.

Proof. 1. The independence of ξ and η yields the independence of $\xi - m_\xi$ and $\eta - m_\eta$, so

$$cov(\xi, \eta) = \mathbb{E}((\xi - m_\xi)(\eta - m_\eta))$$
$$= \mathbb{E}(\xi - m_\xi) \cdot \mathbb{E}(\eta - m_\eta) = 0 \cdot 0 = 0.$$

2. See the proof of the second part of Theorem 4.2.1.

3. This is an immediate consequence of point 5 in Theorem 4.1.1.

4. Point 5 of Theorem 4.1.1 implies that the equality $|cov(\xi, \eta)| = \sigma_\xi \sigma_\eta$ is equivalent to the linear dependence of $\xi - m_\xi$ and $\eta - m_\eta$. Since ξ and η are both not constant, we have $\xi - m_\xi \neq 0$, $\eta - m_\eta \neq 0$, and there exist nonzero α, β with

$$\alpha(\xi - m_\xi) + \beta(\eta - m_\eta) = 0.$$

Clearly

$$\eta = a\xi + b,$$

where $a = -\dfrac{\alpha}{\beta}$, $b = \dfrac{\alpha \mu_\xi}{\beta} + m_\eta$.

Now, from the linearity of the mean, $cov(\xi, \eta) = cov(\xi, a\xi) = a \operatorname{Var}(\xi)$.

5. Multiplying and using the linearity of the mean we have

$$cov(\xi, \eta) = \mathbb{E}((\xi - m_\xi)(\eta - m_\eta))$$
$$= \mathbb{E}(\xi\eta) - m_\xi \mathbb{E}(\eta) - m_\eta \mathbb{E}(\xi) + \mathbb{E}(m_\xi m_\eta)$$
$$= \mathbb{E}(\xi\eta) - m_\xi m_\eta.$$

$\square$

Definition 4.6.2 *The number*

$$\varrho = \varrho(\xi, \eta) = \frac{cov(\xi, \eta)}{\sigma_\xi \sigma_\eta}$$

is called the coefficient of correlation of ξ and η.

The properties of covariance imply immediately the corresponding properties of the coefficient of correlation.

Theorem 4.6.3 *We have:*

1. *If ξ and η independent, then $\varrho(\xi, \eta) = 0$.*
2. *$|\varrho(\xi, \eta)| \leq 1$.*
3. *$|\varrho(\xi, \eta)| = 1$ if and only if $\eta = a\xi + b$, where a, $b \in \mathbb{R}$, with $a \neq 0$, in which case $sign(\varrho) = sign(a)$.*

As we have mentioned before, the coefficient of correlation ϱ may be used as an effective measure of the dependence of random variables. Although the condition $\varrho = 0$, in general, does not imply the independence of ξ and η, in the most important case when the vector (ξ, η) is normally distributed, it does. Then one can say that a ϱ close to zero means very weakly dependent, while $|\varrho|$ close to one means almost linearly related, in which case the sign of ϱ indicates the kind of dependence.

Given $P_{(\xi,\eta)}$ we can compute the covariance and the correlation coefficient. We already know formulas for the means and variances. Besides, Theorems 4.3.2 and 4.3.3 imply, respectively:

In the discrete case:

$$cov(\xi, \eta) = \sum_{i,j} (x_i - m_\xi)(y_j - m_\eta)r_{ij}, \qquad (4.26)$$

or

$$cov(\xi, \eta) = \sum_{i,j} x_i y_j r_{ij} - m_\xi m_\eta. \qquad (4.27)$$

In the continuous case:

$$cov(\xi, \eta) = \int_{\mathbb{R}^2} (x - m_\xi)(y - m_\eta)h(x, y)\, dx dy, \qquad (4.28)$$

or

$$cov(\xi, \eta) = \int_{\mathbb{R}^2} xy\, h(x, y)\, dx dy - m_\xi m_\eta, \qquad (4.29)$$

where x_i, y_j, r_{ij} and $h(x, y)$ determine discrete, respectively continuous, distribution of the random vector (ξ, η) in the standard way.

Example 4.6.1 Again we toss two dice. Let ξ take the value 0 if "6" does not appear on any die and 1 otherwise. Let η be the minimum of the two numbers displayed. We want to find the distribution $P_{(\xi,\eta)}$, the marginal distribution, the means and the variances of both random variables, their covariance and the correlation coefficient. First note that the random variables are dependent. In fact, information about ξ may help in determining η. For example, $\xi = 0$ implies that $\eta \neq 6$. On the other hand, the dependence between ξ and η does not seem to be very strong and surely it is not linear. Thus we may believe that the correlation coefficient is not zero, but is positive (an increase of ξ gives more chance for η to increase) and is essentially < 1.

Clearly, the distribution $P_{(\xi,\eta)}$ is concentrated at points (i, j), $i = 0, 1$, $j = 1, \ldots, 6$. One can easily find all probabilities $P_{(\xi,\eta)}(i, j)$. For example, $P_{(\xi,\eta)}(1, 3)$ means the probability of the event consisting of the elementary events $(3, 6)$, $(6, 3)$, and thus equals $\frac{2}{36}$. We can write down the values of the distribution $P_{(\xi,\eta)}$ in a table:

	1	2	3	4	5	6
0	$\frac{9}{36}$	$\frac{7}{36}$	$\frac{5}{36}$	$\frac{3}{36}$	$\frac{1}{36}$	0
1	$\frac{2}{36}$	$\frac{2}{36}$	$\frac{2}{36}$	$\frac{2}{36}$	$\frac{2}{36}$	$\frac{1}{36}$

Now we can find the marginal distribution.

MAPLE ASSISTANCE 37

We have to input the table into MAPLE.

```
>   for i to 5 do r[0,i] := (11 - 2*i)/36 od:
>   for i to 5 do r[1,i] := 2/36 od:
>   r[0,6] := 0: r[1,6] := 1/36:
```

Compute the distributions of ξ and η.

```
>   for i from 0 to 1 do
>   p[i] := sum(r[i,'j'],'j' =1..6): od:
>   seq(p[i],i=0..1);
```
$$\frac{25}{36}, \frac{11}{36}$$

```
>   for j from 1 to 6 do
>   q[j] := sum(r['i',j],'i' =0..1): od:
>   seq(q[j],j=1..6);
```
$$\frac{11}{36}, \frac{1}{4}, \frac{7}{36}, \frac{5}{36}, \frac{1}{12}, \frac{1}{36}$$

Compute the means and variances.

```
>   m[xi] := sum('i'*p['i'],'i' =0..1);
```
$$m_\xi := \frac{11}{36}$$

```
>   m[eta] := sum('j'*q['j'],'j' =1..6);
```
$$m_\eta := \frac{91}{36}$$

```
>   var[xi] := sum(('i' - m[xi])^2*p['i'],'i' =0..1);
```
$$war_\xi := \frac{275}{1296}$$

```
>   var[eta] := sum(('j'- m[eta])^2*q['j'],'j' =1..6);
```
$$war_\eta := \frac{2555}{1296}$$

Refresh variables i and j.

```
>   i := 'i': j := 'j':
```

Compute the covariance and the corelation coefficient.

```
>   cov := sum(sum(('i'-m[xi])*('j'-m[eta])*r['i','j'],
    'i'= 0..1), 'j' = 1..6);
```
$$cov := \frac{295}{1296}$$

```
>  rho := cov/(var[xi]*var[eta]);
```
$$\rho := \frac{76464}{140525}$$

```
>  evalf(%);
```
$$.5441309376$$

Compute the covariance using the other formula.

```
>  cov2 := sum(sum('i'*'j'*r['i','j'],'i'=0..1),'j' = 1..6)
   - m[xi]*m[eta];
```
$$cov2 := \frac{295}{1296}$$

4.7 MAPLE Session

M 4.1 *Evaluate the mean of a random variable ξ having the discrete distribution*

$$P(\xi = 1) = 0.1, \ P(\xi = 3) = 0.3, \ P(\xi = 5) = 0.1,$$

$$P(\xi = 7) = 0.44, \ P(\xi = 11) = 0.06.$$

```
>  with(stats);
```
[*anova, describe, fit, importdata, random, statevalf, statplots, transform*]
```
>  x := [1,3,5,7,10]: p := [.1,.3,.1,.44,.06]:
   data := [seq(Weight(x[i],p[i]),i = 1..nops(x))];
```

data := [Weight(1, .1), Weight(3, .3), Weight(5, .1), Weight(7, .44),
    Weight(10, .06)]
```
>  evalf(describe[mean](data));
```
$$5.180000000$$

M 4.2 *Find the mean of a random variable ξ taking values*

$$1, \ 3, \ 5, \ 7, \ 10$$

with weights

$$10, \ 30, 10, \ 44, \ 6$$

respectively.

```
>  x := [1,3,5,7,10]: p := [10,30,10,44,6]:
   data := [seq(Weight(x[i],p[i]),i = 1..nops(x))];
```

data := [Weight(1, 10), Weight(3, 30), Weight(5, 10), Weight(7, 44),
    Weight(10, 6)]

```
> evalf(describe[mean](data));
```
$$5.180000000$$

Thus weights are, in this case, more general than the probabilities.

M 4.3 *For the above random variable ξ find its mode, i.e. the value corresponding to the largest probability, and its median $x_{\frac{1}{2}}$ i.e. the value that is at the middle of the sample. Formally, the median is defined by*

$$F(x_{\frac{1}{2}}^{\ -}) \le \frac{1}{2} \le F(x_{\frac{1}{2}}).$$

```
> describe[mode](data);
```
$$7$$
```
> describe[median](data);
```
$$6$$

M 4.4 *A company released data of monthly salaries at its local factory. It appears that 60% earn $2000, 20% earn $2500, 8% earn $2800, 8% earn $6000 and 4% earn $10000 a month. What is the average monthly salary in this factory?*

Obviously, the company has computed the mean.
```
> x := [2000,2500,2800,6000,10000]: p := [60,20,8,8,4]:
  data := [seq(Weight(x[i],p[i]),i = 1..nops(x))];
```

$$data := [\text{Weight}(2000, 60), \text{Weight}(2500, 20), \text{Weight}(2800, 8),$$
$$\text{Weight}(6000, 8), \text{Weight}(10000, 4)]$$
```
> evalf(describe[mean](data));
```
$$2804.$$

Trade unions have insisted that the mode, or at least the median, should be used as a measure of the central tendency.
```
> evalf(describe[mode](data));
```
$$2000.$$
```
> evalf(describe[median](data));
```
$$2000.$$

A compromise has been reached at last ....
```
> data2 := [seq(Weight(x[i],p[i]),i = 1..nops(x)-2)];
```
$$data2 := [\text{Weight}(2000, 60), \text{Weight}(2500, 20), \text{Weight}(2800, 8)]$$
```
> evalf(describe[mean](data2));
```
$$2186.363636$$

However, another independent trade union is performing its own investigation. First, they choose 10 employees according to the official distribution.

```
> prob := evalf(p/100,1);
```
$$prob := [.6, .2, .08, .08, .04]$$
```
> salary := empirical[op(prob)]:
> stats[random,salary](10);
```
$$1.0, 1.0, 2.0, 1.0, 1.0, 1.0, 1.0, 1.0, 2.0, 1.0$$
```
> sal := i ->x[i]:
> results := map(sal@trunc,[%%]);
```
$$results := [2000, 2000, 2500, 2000, 2000, 2000, 2000, 2000, 2500, 2000]$$

And now they compute an average salary on their own . . . .
```
> describe[mean](results);
```
$$2100$$

M 4.5 *A population of $N = 1000$ individuals has to be investigated for the presence of a dangerous virus. Because the analysis of blood is expensive, one divides the population into small groups and performs a joint analysis for each group, and then, if the virus is detected in a particular group, an extra analysis is performed for all individuals in this group. Since the probability of being infected by the virus is small, say $p = 0.01$, this seems reasonable.*

Perform a simulation of the investigation, assuming a group size $= 30$. Is this size optimal?

First, we simulate the investigation process.
```
> infect := empirical[0.99,0.01]:
> N := 1000: groupsize := 30:
  K := iquo(N,groupsize); r := irem(N,groupsize);
```
$$K := 33$$
$$r := 10$$
```
> anal := 0:
  for i from 1 to K do
  agroup[i] := [stats[random,infect](groupsize)]:
  if member(2.0,agroup[i]) then anal := anal + groupsize
  fi:
  anal := anal + 1
  od:
> if r > 0 then
  rgroup := [stats[random,infect](r)]:
  if member(2.0,rgroup) then anal := anal + r fi:
  anal := anal + 1:
  fi:
> anal;
```

Next, we will find the expected value of the number of analyses needed for an arbitrary group size. However, for simplicity we assume that the group size divides the population size.

Let $q = 1 - p$. In a single group of size n we may have 1 or $n + 1$ analyses. The mean and the variance for any group is then

```
>  Eg := q^n + (n +1)*(1 - q^n):
>  Vg := (1 -Eg)^2*q^n + (n+1 - Eg)^2*(1 -q^n):
```

By a property of the mean and by independence we have for the whole population.

```
>  Etot := N/n*Eg: std := sqrt((N/n)*Vg):
```

By our assumption:

```
>  q := 0.99:
```

We plot the expected value of the number of analyes as a function of the group size. Taking into account remarks made after the proof of Chebyshev's Inequality (see page 99) we also plot the function which indicates the bound guaranteeing with probability $\frac{8}{9}$ the number of analyses needed.

```
>  plot([Etot,Etot+3*std], n = 1..50,color=[black,blue],
     style=[line,point],thickness=[2,1]);
```

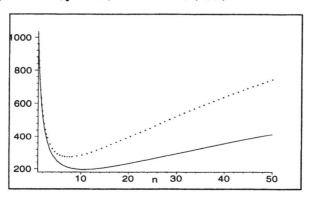

This plot suggests that a group size equal to 10 would be optimal. Have a look at the particular values.

```
>  eval(Etot,n=10),eval(Etot+3*std,n=10);
            195.6179250, 195.6179250 + 27.89760272 √10
```

We should expect 196 analyses and we are actually sure that this number will be less than 300.

Let us check the situation when the probability of infection is large.

```
>  q := 0.9:
>  plot([Etot,Etot+3*std], n = 1..25,color=[black,blue],
     style=[line,point],thickness=[2,1]);
```

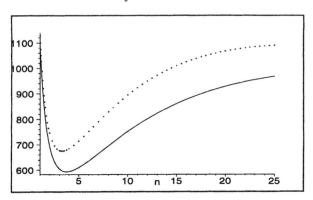

M 4.6 *Let ξ be a random variable taking values $1, 2, 3, \ldots, 100$ with equal probabilities $\frac{1}{100}$ and let $\eta = a\xi^2 + b\xi + c + d\varepsilon$, where a, b, c, d are constants and ε is a random variable uniformly distributed on the interval $(0, 10)$. Compute the correlation coefficient of ξ and η and plot 100 values of the random vector (ξ, η) for three particular sets of parameters.*

We build a procedure to be used three times.

```
>   corr := proc(a,b,c,d)
    global graph, rho:
    local K,xi,eta,x:
>   K := [seq([x,a*x^2+b*x+c+d*10^(-11)*rand()], x=1..100)]:
>   graph := plot(K, 0..100,
    style=point,symbol=CIRCLE, color= black ):
>   xi := map2(op,1,K): eta := map2(op,2,K):
>   rho := stats[describe,linearcorrelation](xi,eta):
>   end:
```

We allow randomness and exclude trends.

```
>   corr(0,0,40,10):
>   evalf(rho);
>   graph;
```
.03370012119

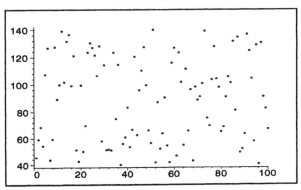

We admit linear tendency with small randomness.

```
>  corr(0,0.3,40,1):
>  evalf(rho);
>  graph;
```

.9471809724

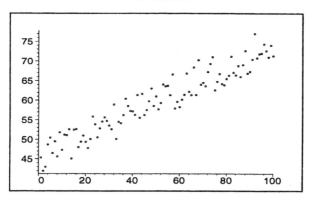

Finally, we include a quadratic term.

```
>  corr(0.05,-6,200,5):
>  evalf(rho);
>  graph;
```

−.5390036706

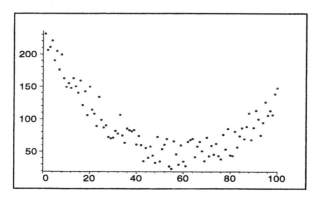

4.8 Exercises

E 4.1 A calculating burglar is trying to open a safe. He knows that exactly one key among the 20 he has is good. How long can he expect to work, if every next key is chosen by chance?

E 4.2 From a box containing 5 red, 4 white and 3 green balls, pairs of balls are randomly taken out and then put back. How long, on average, has one to wait for a pair of the same color? Consider two cases: One takes out balls to complete each pair (a) with replacement, (b) without replacement.

E 4.3 What fee should Mr. Smith require for a single game in Example 4.1.1 if he allows players to toss his die as long as they obtain a number greater than 1?

   Hint. There are at least two strategies for solving this problem. One of them is almost trivial.

E 4.4 How long does one have to toss a die to get "6" with at least probability 0.96?

E 4.5 How long does one have to toss a die to get its all faces with at least probability $\frac{8}{9}$?

E 4.6 Let a_n denote the minimum number appearing on a die in the series of n tosses. Predict and then compute $\lim_{n\to\infty} \mathbb{E}(a_n)$ and $\lim_{n\to\infty} \mathrm{Var}\,(a_n)$.

E 4.7 What is the expected area of the rectangle for which the lengths of the sides are randomly chosen from the interval $(0,4)$?

E 4.8 Assuming that the birth of a boy and the birth of a girl have equal probability, find the probability that among N newborn the numbers of boys and girls differ more than 10% of N. Give an explicit answer for $N = 10$, $N = 20$ and $N = 100$. See also Exercise 5.13.

E 4.9 Let ξ be a random variable and $x_0 \in \mathbb{R}$. Let $a = P(\xi \le x_0)$. For an i.i.d. sequence $\xi_1, \xi_2, \xi_3, \ldots$ such that $P_{\xi_i} = P_\xi$ for each i, define quantities

$$A_n = \frac{1}{n}\#\{k \le n : \xi_k \le x_0\}, \quad n = 1, 2, 3, \ldots.$$

This quantity is called the empirical distribution function and is an example of the so called estimator of the parameter a. Prove that

1. A_n are random variables.
2. $\mathbb{E}(A_n) = a$, for all n. (One says that the estimator is unbiased.)
3. $A_n \xrightarrow{P} a$. (One says that the estimator is (weakly) consistent.)
4. $A_n \xrightarrow{1} a$. (One says that the estimator is (strongly) consistent.)

   Hint. A_n is the arithmetic average of the indicators of sets $\{\xi_i \le x_0\}$.

E 4.10 Let ξ be a random variable with finite mean m and variance σ^2. For an i.i.d. sequence $\xi_1, \xi_2, \xi_3, \ldots$ such that $P_{\xi_i} = P_\xi$ for each i, define the estimators

$$\bar\xi_n = \frac{1}{n}\sum_{i=1}^{n} \xi_i,$$

$$s_n = \frac{1}{n-1} \sum_{i=1}^{n} (\xi_i - \bar{\xi}_n)^2$$

and

$$s_n^* = \frac{1}{n} \sum_{i=1}^{n} (\xi_i - m)^2.$$

Prove that $\bar{\xi}_n$ is an unbiased and consistent estimator of m, while both s_n and s_n^* are unbiased and consistent estimators of σ^2.

E 4.11 In Example 4.5.1 determine the distribution functions of variables ξ_{kl} and then show that the sequence ξ_{kl} converges in distribution to zero.

E 4.12 Find the correlation coefficient of the minimum and maximum of pips on two randomly tossed dice.

E 4.13 Let the random vector (ξ, η) is uniformly distributed on the triangle with vertices $(-1,0)$, $(1,0)$ $(0,1)$. Find the marginal distributions. Find the correlation coefficient.

5 A Tour of Important Distributions

This chapter reviews a selection of the best known probability distributions, which are important from both a theoretical and a practical point of view. Two other probability distributions, needed for statistical inference, will be mentioned in Chapter 6.

5.1 Counting

How many experiments will be successful? How often will "our" numbers in a game of chance be good? How many emergency calls on average will the police receive on a working day? How many deaths in street accidents can be expected next summer? These and similar questions can be answered if one knows the nature of the phenomena, or more precisely, an appropriate probability distribution. It seems that many different processes run according to the same pattern, for example, a number of real processes are nothing else than drawing with or without replacement. Below we review some typical probability distributions that are usually shared by random variables associated with processes in which we want to count the appearance or occurrence of some event.

5.1.1 The Binomial Distribution

A probability distribution P is called a *binomial distribution* if there are numbers p and q with $0 < p, q < 1$, $p + q = 1$ and an integer $n > 0$ such that

$$P(k) = \binom{n}{k} p^k q^{n-k} \quad \text{for } k = 0, 1, \ldots, n.$$

MAPLE ASSISTANCE 38

We plot an example of the binomial distribution.

```
> restart:
> with(plottools):
> g := x -> stats[statevalf,pf,binomiald[12,0.6]](x):
> a := {seq(curve([[i,0],[i,g(i)]],thickness=2),i=0..12)}:
  plots[display](a,xtickmarks=6,ytickmarks=2,
  view = [0..12.5,0..0.3]);
```

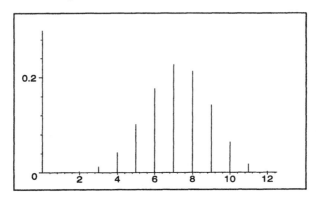

We can easily check using the binomial formula of Newton that $\sum_{k=0}^{n} P(k) = 1$, so the above condition determines a probability distribution. We assume that the reader already knows some examples where this distribution appears. We also refer to Example 1.5.6 where we derived a formula for the probability of occurrence of k successes in n Bernoulli trials. Using the language of random variables we can reformulate Example 1.5.6 as follows.

Theorem 5.1.1 *Let $\xi_1, \ldots, \xi_n$ be i.i.d. variables having a common two-point distribution (see page 65):*

$$P(\xi_i = 0) = q, \qquad P(\xi_i = 1) = p, \quad for\ i = 1, \ldots, n.$$

where $p, q > 0$ with $p + q = 1$. (Such random variables are called Bernoulli trials.) Then the sum $S_n = \xi_1 + \ldots + \xi_n$ has a binomial distribution.

Proof. Arguments are similar to those in the proof of Example 1.5.6, so we omit the details. $\square$

Example 5.1.1 *Drawing with replacement*
Assume that a population consists of N elements and exactly K of them share some property, say, $\mathcal{W}$. Write $p = K/N$. We draw with replacement k elements and obtain ξ elements with property $\mathcal{W}$. Then ξ has a binomial distribution.

Theorem 5.1.2 *The expectation and variance of a random variable having a binomial distribution are given by*

$$\mathbb{E}(\xi) = np, \qquad\qquad \mathrm{Var}\,(\xi) = npq. \qquad\qquad (5.1)$$

Proof. As mentioned on page 92, the moments of a random variable are uniquely determined by its probability distribution. Hence we can compute .the expectation and the variance for a random variable having the same distribution as ξ, and the result will apply to ξ. An appropriate random variable for us is the sum S_n in Theorem 5.1.1. By Example 4.3.2 we have

$\mathbb{E}(\xi_i) = p$ and $\text{Var}(\xi_i) = pq$, for $i = 1, \ldots, n$, so by the additivity property of expectation (see Point 4 in Theorem 4.1.1) we have

$$\mathbb{E}(S_n) = \sum_{i=1}^{n} \mathbb{E}(\xi_i) = \sum_{i=1}^{n} p = np.$$

In addition, the additivity property of variance (Point 2 in Theorem 4.2.1) applies here because the random variables ξ_i are independent, so

$$\text{Var}(S_n) = \sum_{i=1}^{n} \text{Var}(\xi_i) = \sum_{i=1}^{n} pq = npq.$$

□

5.1.2 The Multinomial Distribution

A generalisation of the binomial distribution is the *multinomial distribution*.

A probability distribution P is called a multinomial distribution if there exists natural numbers n and r, real numbers $p_i > 0$, $i = 1 \ldots r$, where $r > 1$, with $\sum_{i=1}^{r} p_i = 1$ such that for any collection of nonnegative integers $k_1, \ldots, k_r$ with $\sum_{i=1}^{r} k_i = n$ we have

$$P(k_1, \ldots, k_r) = \frac{n!}{k_1! \cdot \ldots \cdot k_r!} \, p_1^{k_1} \cdot \ldots \cdot p_r^{k_r}.$$

For $r = 2$ the multinomial distribution is just the binomial distribution. (We just put $p_1 = p$ and $p_2 = q$). Imagine some experiment to be repeated n times according to the following rules.

1. Every time there are exactly r different outomes of the experiment, say "1", ..., "r".
2. The probability of a given outome is fixed in all experiments. Denote these probabilities as p_i, $i = 1 \ldots r$.
3. The experiments are run independently.

Let $\xi_1, \ldots, \xi_r$ denote the number of experiments resulting in outcomes "1", ..., "r", respectively. Then by a simple induction argument one can see that the random vector $(\xi, \ldots, \xi_r)$ has the multinomial distribution.

Example 5.1.2 Five tests for a student class consisting of four groups of 15, 11, 10 and 14 students are conducted and evaluated by the assistants during the semester. The professor selects one paper from each test to check it himself. What is the probability that the professor will see papers coming from all groups.

We consider the multinomial distribution here with parameters $n = 5$, $r = 4$, $p_1 = \frac{15}{50}$, $p_2 = \frac{11}{50}$, $p_3 = \frac{10}{50}$, $p_4 = \frac{14}{50}$. We want to compute the following:

$$P(2,1,1,1) + P(1,2,1,1) + P(1,1,2,1) + P(1,1,1,2).$$

Key in the required parameters and define a procedure P for computing the particular probabilities.

```
> n := 5: r := 4: p1,p2,p3,p4 := 15/50, 11/50, 10/50, 14/50;
```
$$p1,\ p2,\ p3,\ p4 := \frac{3}{10},\ \frac{11}{50},\ \frac{1}{5},\ \frac{7}{25}$$

```
> P := proc(k1,k2,k3,k4)
    combinat[multinomial](n,k1,k2,k3,k4)*
    p1^k1*p2^k2*p3^k3*p4^k4
    end;
```

$P := \mathbf{proc}(k1,\ k2,\ k3,\ k4)$

$\quad combinat_{multinomial}(n,\ k1,\ k2,\ k3,\ k4) \times p1^{k1} \times p2^{k2} \times p3^{k3} \times p4^{k4}$

end

Answer the question:

```
> P(2,1,1,1) + P(1,2,1,1) +P(1,1,2,1) +P(1,1,1,2);
```
$$\frac{693}{3125}$$

```
> evalf(%);
```
$$.2217600000$$

5.1.3 The Poisson Distribution

A probability distribution P is called a *Poisson distribution* if there exists a number $\lambda > 0$ such that

$$P(k) = e^{-\lambda}\frac{\lambda^k}{k!}, \quad \text{for } k = 0, 1, 2, \dots$$

We plot an example of the Poisson distribution.

```
> g := x -> stats[statevalf,pf,poisson[5]](x):
  a := {seq(curve([[i,0],[i,g(i)]],thickness=2),i=0..12)}:
  plots[display](a,xtickmarks=6,ytickmarks=2,
  view = [0..12.5,0..0.2]);
```

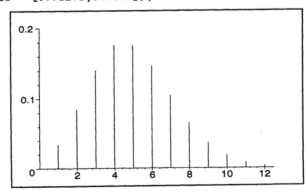

Many phenomena seem to occur according to a Poisson distribution. The following theorem gives to some extent an explanation. Namely, one can think about a Poisson distribution as if it were a binomial distribution with large n, very small p and moderate product np denoted by λ, typically with $0.1 \leq \lambda = np \leq 10$ in practice. In other words, the Poisson distribution describes the number of successes in a great number of independent experiments having almost zero probability of success of a single trial. This fact has also been confirmed in a lot of cases and the compatibility of the Poisson distribution with appropriate λ to real data is remarkably good. We establish our result in the simplest possible form.

Theorem 5.1.3 *Let $p_n > 0$ be such a sequence of numbers that $\lim_{n\to\infty} np_n = \lambda > 0$. Let k be a nonnegative integer. Then*

$$\lim_{n\to\infty} \binom{n}{k} p_n^k (1 - p_n)^{n-k} = e^{-\lambda} \frac{\lambda^k}{k!}$$

Proof. Putting $\lambda_n = np_n$ we have

$$\binom{n}{k} p_n^k (1-p_n)^{n-k} = \frac{\lambda_n^k}{k!} \cdot \frac{n(n-1)\cdot \ldots \cdot (n-k+1)}{n^k} \cdot (1 - \frac{\lambda_n}{n})^n \cdot (1 - \frac{\lambda_n}{n})^{-k}.$$

Since k is fixed, the last factor tends to 1. The second factor equals $1 \cdot (1 - \frac{1}{n}) \cdot \ldots \cdot (1 - \frac{k+1}{n})$, so it also tends to 1. Finally, the first factor and the third one tend to $\frac{\lambda^k}{k!}$ and $e^{-\lambda}$, respectively.   □

Theorem 5.1.4 *The expectation and variance of a random variable having the Poisson distribution are given by*

$$\mathbb{E}(\xi) = \lambda, \qquad\qquad \mathrm{Var}\,(\xi) = \lambda. \qquad\qquad (5.2)$$

Proof. Instead of a standard proof we ask MAPLE to do some symbolic computations:

MAPLE ASSISTANCE 41

```
>  g := k->exp(-lambda)*lambda^k/k!;
```
$$g := k \rightarrow \frac{e^{(-\lambda)} \lambda^k}{k!}$$
```
>  E := sum(k*g(k),k=0..infinity);
```
$$E := e^{(-\lambda)} \lambda e^{\lambda}$$
```
>  E := simplify(E);
```
$$E := \lambda$$
```
>  Var := simplify(sum(k^2*g(k),k=0..infinity) - E^2);
```
$$Var := \lambda$$

In the following table we compare two pairs of binomial and Poisson distributions with appropriate parameters n, p and $\lambda = np$.

Binomial versus Poisson distribution

k	n = 100, p = 0.01 binomial Poisson distrib. distrib.		n = 50, p = 0.1 binomial Poisson distrib. distrib.		n = 100, p = 0.1 binomial Poisson distrib. distr.	
0	0.3660	0.3679	0.0052	0.0067	0.0000	0.0000
1	0.3697	0.3679	0.0286	0.0337	0.0003	0.0005
2	0.1849	0.1839	0.0779	0.0842	0.0016	0.0023
3	0.0610	0.0613	0.1386	0.1404	0.0059	0.0076
4	0.0149	0.0153	0.1809	0.1755	0.0159	0.0189
5	0.0029	0.0031	0.1849	0.1755	0.0339	0.0378
6	0.0005	0.0005	0.1541	0.1462	0.0596	0.0631
7	0.0001	0.0001	0.1076	0.1044	0.0889	0.0901
8	0.0000	0.0000	0.0643	0.0653	0.1148	0.1126
9	0.0000	0.0000	0.0333	0.0363	0.1304	0.1251
10	0.0000	0.0000	0.0152	0.0181	0.1319	0.1251
11	0.0000	0.0000	0.0061	0.0082	0.1199	0.1137
12	0.0000	0.0000	0.0022	0.0034	0.0988	0.0948
13	0.0000	0.0000	0.0007	0.0013	0.0743	0.0729
14	0.0000	0.0000	0.0002	0.0005	0.0513	0.0521
15	0.0000	0.0000	0.0001	0.0002	0.0327	0.0347

Let us interpret the middle two columns graphically.

MAPLE ASSISTANCE 42

```
>  g := x -> stats[statevalf,pf,binomiald[50,0.1]](x):
   h := x -> stats[statevalf,pf,poisson[5]](x):
>  bd := plots[pointplot]([seq([i,g(i)],i = 0..15)],
   color = black, symbol = cross):
   pd := plots[pointplot]([seq([i,h(i)],i = 0..15)],
   color = black, symbol = circle):
>  plots[display](bd,pd);
```

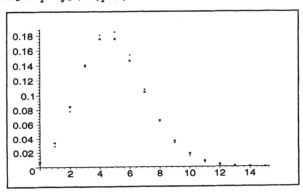

5.1.4 The Hypergeometric Distribution

The binomial, the multinomial and the Poisson distributions discussed above arise when one wants to count the number of successes in situations which, generally speaking, correspond to drawing with replacement. The following distribution appears when we consider drawing without replacement.

A probability distribution P is called a *hypergeometric distribution* if there exist positive integers N, n and positive numbers p, q, $p + q = 1$ such that for any $k = 0, 1, 2, \ldots n$ we have

$$P(k) = \frac{\binom{Np}{k} \binom{Nq}{n-k}}{\binom{N}{k}}.$$

Here we are using a generalized Newton symbol (generally, Np is not a positive integer), which is defined for real x and nonnegative integer k by

$$\binom{x}{k} = \frac{x(x-1)\ldots(x-k+1)}{k!}.$$

MAPLE ASSISTANCE 43

We plot an example of the hypergeometric distribution. MAPLE supports this distribution in the following meaning explained under "Help":

- The hypergeometric[N1, N2, n], with N1 equal to the size of the success population, N2 equal to the size of the failure population and n equal to the sample size, has the probability density function

binomial(N1,x)*binomial(N2,n-x)/binomial(N1+N2,n).

Constraints: n<=N1+N2.

Comparing with our notation we have
$N = N1 + N2$ and $p = \dfrac{N1}{N1 + N2}$, $q = \dfrac{N2}{N1 + N2}$, with the same n.

Fix some parameters and plot the appropriate graph:

```
> N1 := 20: N2 := 30: n := 5:
> g := x -> stats[statevalf,pf,hypergeometric[N1,N2,n]](x):
  a := {seq(curve([[i,0],[i,g(i)]],thickness=2),i=0..n)}:
  plots[display](a,xtickmarks = 6 ,ytickmarks = 2,
  view = [0..5.5, 0..0.4]);
```

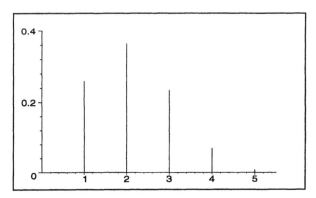

Example 5.1.3 *Drawing without replacement*
Assume that a population consists of N elements. Let p be the probability that any element from this population shares some property, say A. We draw without replacement n elements and denote by ξ the number of those with the property A. Applying the same reasoning as in Example 1.2.1, we can show that ξ has a hypergeometric distribution.

One can also prove the following:

Theorem 5.1.5 *The expectation and variance of a random variable having hypergeometric distribution are given by*

$$\mathbb{E}(\xi) = np, \qquad \qquad \text{Var}\,(\xi) = npq\frac{N-n}{N-1}. \qquad (5.3)$$

From the above formulas and from formulas (5.1) we immediately have the following.

Remark 5.1.1 *Drawing n elements with or without replacement from a population of N elements containing a determined fraction we get **on average the same** result for that fraction, though drawing without replacement results in less variance. On the other hand, if the size of the population N is large enough compared with the sample size n, the variances in both cases are almost equal. In fact, both distributions are then almost identical.*

5.2 Waiting Times

How long must one toss a die to get "6"? How long is the time interval between two consecutive calls arriving at a telephone exchange? How many street accidents occur during a working day? Like in the preceding Section we review a few typical distributions considered mostly in connection with waiting times.

5.2.1 The Geometric Distribution

A probability distribution P is called a *geometric distribution* if there exists numbers p and q with $0 < p, q < 1$ and $p + q = 1$ such that

$$P(k) = q^{k-1}p \text{ for } k = 1, 2, 3, \ldots.$$

MAPLE ASSISTANCE 44

We plot an example of the geometric distribution.

```
>  p := 1/4:
>  g := x -> (1-p)^(x-1)*p:
   a := {seq(curve([[i,0],[i,g(i)]],thickness=2),i=1..12)}:
   plots[display](a, xtickmarks = 6, ytickmarks = 2,
   view = [0..12.5, 0..0.3]);
```

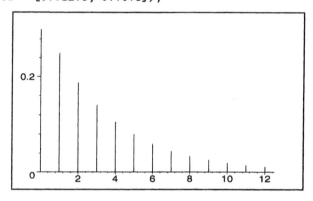

The geometric distribution is closely related to a sequence of Bernoulli trials. The following theorem says that the waiting time for the first success in such a sequence of trials has a geometric distribution. A special case of waiting time, waiting time for the first "6" while tossing a die, was considered in Example 1.2.3.

Theorem 5.2.1 *Let $\xi_1, \xi_2, \xi_3, \ldots$ be independent random variables with a common two-point distribution. Then the function*

$$T = \min\{n \geq 1 : \xi_n = 1\}$$

is a random variable with a geometric distribution.

Proof. We note that the event $\{T = n\}$ is just the same as the event $\{\xi_1 = 0, \ldots, \xi_{n-1} = 0, \xi_n = 1\}$. Independence of the ξ_i yields

$$\begin{aligned}
P(T = n) &= P(\xi_1 = 0, \ldots, \xi_{n-1} = 0, \xi_n = 1) \\
&= P(\xi_1 = 0) \cdot \ldots \cdot P(\xi_{n-1} = 0) \cdot P(\xi_n = 1) \\
&= q^{n-1}p.
\end{aligned}$$

$\square$

Theorem 5.2.2 *The expectation and variance of a random variable having the geometric distribution are given by*

$$\mathbb{E}(\xi) = \frac{1}{p}, \qquad \text{Var}(\xi) = \frac{q}{p^2}. \qquad (5.4)$$

Again we use the MAPLE symbolic computation facility:

MAPLE ASSISTANCE 45

```
> p := 'p':
> E := sum(k*g(k),k=1..infinity);
> Var := simplify(sum(k^2*g(k),k=1..infinity) - E^2);
```

$$E := \frac{1}{p}$$

$$Var := -\frac{-1+p}{p^2}$$

Consider a special case:
```
> subs(p=1/4,E), subs(p=1/4,Var);
```
$$4, 12$$

5.2.2 The Negative Binomial Distribution

A probability distribution P is called a *negative binomial distribution* (or sometimes a *Pascal distribution*) if there exists an integer $r \geq 1$ and a real number $p \in (0,1)$ such that

$$P(r+k) = \binom{r+k-1}{r-1} p^r (1-p)^k \text{ for } k = 0,1,2,\ldots.$$

We note that the geometric distribution is a special case of the negative binomial distribution.

MAPLE ASSISTANCE 46

We plot an example of the negative binomial distribution.
```
> r := 'r': p := 'p':
> g := k->binomial(r+k-1,r-1)*p^r*(1-p)^k;
```
$$g := k \to \text{binomial}(r+k-1, r-1) \, p^r \, (1-p)^k$$
```
> r := 5: p := 1/4:
> a := {seq(curve([[r+k,0],[r+k,g(k)]],thickness=2),k=0..30)
}:
  plots[display](a, xtickmarks = 6, ytickmarks = 2,
  view = [0..36.5, 0..0.07]);
```

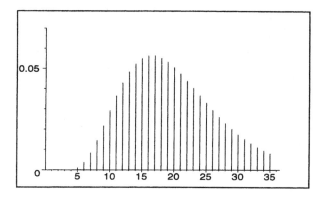

Theorem 5.2.3 *Let $\xi_1, \xi_2, \xi_3, \ldots$ be a sequence of independent Bernoulli trials with p as the common probability of success. Define:*

$$T_r := \min \{n : \exists 1 \leq k_1 < \ldots < k_r = n \text{ such that } \xi_{k_i} = 1 \text{ for } i = 1, \ldots, r\}.$$

Then T_r is a random variable with a negative binomial distribution.

In other words, the waiting time for the first r successes in an infinite Bernoulli scheme has a negative binomial distribution.

Proof. A standard proof is similar to the one of Theorem 5.2.1.   □

One can also prove a theorem that describes the problem of waiting times from another point of view.

Theorem 5.2.4 *Let $T_1, \ldots, T_r$ be a sequence of independent random variables with a common geometric distribution with parameter p.*

Then the sum $T_1 + \ldots + T_r$ has the negative binomial distribution with parameters r and p.

Theorem 5.2.5 *The expectation and variance of a random variable having the negative binomial distribution are given by*

$$\mathbb{E}(\xi) = \frac{r}{p}, \qquad \text{Var}\,(\xi) = \frac{r(1-p)}{p^2}. \tag{5.5}$$

Proof. It follows from the previous theorem and formulas (5.4) for the expectation and the variance in the geometric distribution.   □

5.2.3 The Exponential Distribution

A probability distribution P is called an exponential distribution if its density function is

$$f(x) = \begin{cases} 0 & \text{for } x < 0, \\ \lambda e^{-\lambda x} & \text{for } x \geq 0, \end{cases}$$

where $\lambda > 0$ is a parameter.

MAPLE ASSISTANCE 47

We plot an example of the density of the exponential distribution.

```
>   lambda := 0.25:
>   f := x->piecewise(0 <= x,lambda*exp(-lambda*x));
              f := x → piecewise(0 ≤ x, λe^(-λx))
>   plot(f,-1..20, discont = true, thickness = 2,
      color = BLACK);
```

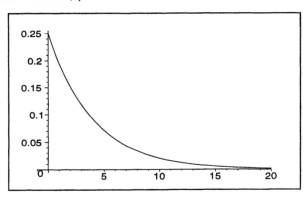

From (3.6) we can compute the distribution function of the exponential distribution. After a standard integration we obtain

$$F(x) = \int_{-\infty}^{x} f(t)\, dt = \begin{cases} 0 & \text{for } x < 0 \\ 1 - e^{-\lambda x} & \text{for } x \geq 0. \end{cases}$$

The plot of the geometric distribution on page 129 and the above plot of a density of the exponential distribution suggest that there is a connection between these two distributions. This is the case, and the following reasoning will partly explain it. In fact, we will show that the exponential distribution is a continuous counterpart of the geometric distribution.

We demonstrate, heuristically, that the waiting time for the first success in an infinite Bernoulli scheme has the exponential distribution with parameter λ, provided there is a very small time between successive trials and the probability of success in each trial is proportional to this time with λ as the proportionality coefficient. To be more precise, let us fix some notation. Let $\lambda > 0$ be fixed. Given $\delta > 0$ set $p = p_\delta = \lambda\delta$. Let $\xi_1, \xi_2, \xi_3, \ldots$ be a sequence of independent random variables with a common two-point distribution with p as the probability of success, and define

$$T_\delta = \delta \cdot \min\{n \geq 1 : \xi_n = 1\}.$$

Let F be the exponential distribution function with parameter λ.

Theorem 5.2.6 *For any* $t \in \mathbf{R}$

$$F_{T_\delta}(t) \longrightarrow F(t) \quad when \quad \delta \longrightarrow 0.$$

Proof. The case $t \leq 0$ is trivial, so let $t > 0$. Taking into account that the random variable $\dfrac{T_\delta}{\delta}$ has a geometric distribution (see Theorem 5.2.1) and denoting the integer part of $\dfrac{t}{\delta}$ by n, we have

$$
\begin{aligned}
F_{T_\delta}(t) &= P(T_\delta \leq t) = 1 - P(T_\delta > t) \\
&= 1 - P\left(\frac{T_\delta}{\delta} > \frac{t}{\delta}\right) \\
&= 1 - \sum_{k=n+1}^{\infty} (1-p)^{k-1}p \\
&= 1 - (1-p)^n \\
&= 1 - \left(1 - \frac{\lambda}{\delta^{-1}}\right)^{\delta^{-1}t - r_\delta} \\
&\longrightarrow 1 - e^{-\lambda t} = F(t),
\end{aligned}
$$

for $\delta \to 0$, as $0 \leq r_\delta = \frac{t}{\delta} - n < 1$. $\qquad\qquad\square$

Theorem 5.2.7 *The expectation and variance of a random variable having exponential distribution are*

$$\mathbb{E}(\xi) = \frac{1}{\lambda}, \qquad\qquad \mathrm{Var}\,(\xi) = \frac{1}{\lambda^2}. \qquad\qquad (5.6)$$

We call MAPLE to compute the appropriate integrals.

MAPLE ASSISTANCE 48

We first remove the numeric value from λ.

```
>  lambda := 'lambda':
```

Now, we can compute expectation and variance by applying formula (4.16) for the expectation and formula (4.4) for the variance, respectively, remembering that $f(x) = 0$ for negative x.

```
>  E := int(x*f(x), x = 0..infinity);
```

$$
E := \lim_{x \to \infty}
\begin{cases}
0 & x \leq 0, \\[2mm]
-\dfrac{x\,e^{(-\lambda x)}\lambda + e^{(-\lambda x)} - 1}{\lambda} & 0 < x,
\end{cases}
$$

This (correct) result looks a bit strange. Namely, MAPLE does not know what sign the parameter λ has.

```
>  assume(lambda > 0):
>  E := int(x*f(x), x = 0..infinity);
```

$$E := \frac{1}{\lambda^{\tilde{}}}$$

```
>  Var := int(x^2*f(x), x = 0..infinity) - E^2;
```

$$Var := \frac{1}{\lambda^{\sim 2}}$$

5.2.4 The Erlang Distribution

A probability distribution P is called an *Erlang distribution* if its density function is

$$f_n(x) = \begin{cases} 0 & \text{for } x < 0 \\ \dfrac{\lambda(\lambda x)^{n-1}}{(n-1)!} e^{-\lambda x} & \text{for } x \geq 0 \end{cases}$$

where n is a positive integer and $\lambda > 0$.

The Erlang distribution is a continuous counterpart of the negative binomial distribution. The following theorem shows the relation between Erlang and exponential distributions.

Theorem 5.2.8 *Let $T_1, \ldots, T_n$ be independent random variables with a common exponential distribution with parameter λ. Let $S_n = T_1 + \ldots + T_n$.*

Then S_n has Erlang distribution with parameters n and λ.

Proof. We omit a detailed proof. It can be done by using formula (3.15) and mathematical induction. □

The above Theorem immediately implies the following theorem.

Theorem 5.2.9 *The expectation and variance of a random variable having an Erlang distribution are*

$$\mathbb{E}(\xi) = \frac{n}{\lambda}, \qquad \text{Var}(\xi) = \frac{n}{\lambda^2}. \qquad (5.7)$$

We will plot a collection of Erlang distribution density functions.

MAPLE ASSISTANCE 49

```
>  k := 10: lambda := 0.25:
>  for n from 2 to k do
   f[n] := x->piecewise(0 <=
   x,exp(-lambda*x)*lambda*(lambda*x)^(n-1)/(n-1)!)
   od:
>  gr.($2..k) := seq(plots[display](plot(f[n],0..40,
   discont = true, color=BLACK, xtickmarks = [20,40],
   ytickmarks = [0,0.02], thickness = 2)), n=2..k):
   plots[display](gr.($2..k),insequence=true):
```

Using ";" instead of ":" in the above command we can define a plot that can be animated with a mouse.

```
>  plots[display](%);
```

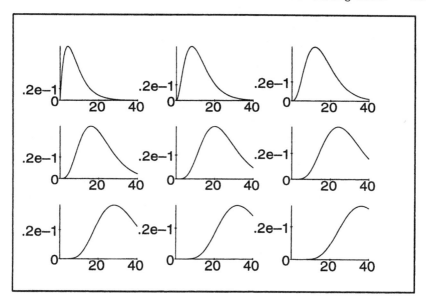

5.2.5 The Poisson Process

We now establish a theorem that explains the deep connection between the exponential and Poisson distributions. We will do this in terms of a stochastic process, i.e. a family $\{N_t\}$ of random variables parametrized by positive instants of time t.

We give an example of a Poisson process, see Definition 7.4.1 on page 212. Suppose, roughly speaking, that N_t denotes the number of successes in an infinite Bernoulli scheme, provided the trials are repeated very often and the probability of success on a small time interval Δt approximately equals $\lambda \Delta t$. Thus we have the situation established in Theorem 5.2.6 and the disscussion preceding it. Then, the waiting time for the first success has an exponential distribution with parameter λ, and hence the waiting time for the first n successes has an Erlang distribution (see Theorem 5.2.8). Finally, the following Theorem guarantees that N_t has the Poisson distribution, so the family $\{N_t\}$ is a Poisson process.

Theorem 5.2.10 *Let $T_1, T_2, T_3, \ldots$ be independent random variables with common exponential distribution with parameter λ. Let $S_n = T_1 + \ldots + T_n$ with $S_0 = 0$. Define*
$$N_t := max\,\{n : S_n \leq t\},$$
for each $t > 0$.

Then N_t has a Poisson distribution with parameter λt.

Proof. Note that the event $\{N_t = k\}$ is the same as the event $\{S_k \leq t\} \setminus \{S_{k+1} \leq t\}$, so

$$P(N_t = k) = F_k(t) - F_{k+1}(t),$$

where F_k is the distribution function of S_k. From Theorem 5.2.8 we know that S_t is Erlang distributed. Hence

$$F_k(t) = \int_0^t \frac{\lambda(\lambda x)^{k-1}}{(k-1)!} e^{-\lambda x} \, dx \quad \text{for } t > 0.$$

By induction one can show that

$$F_k(t) = 1 - e^{-\lambda t} \left(1 + \frac{\lambda t}{1!} + \ldots + \frac{(\lambda t)^{n-1}}{(n-1)!} \right),$$

so $P(N_t = k) = \dfrac{(\lambda t)^k}{k!} e^{-\lambda t}.$ □

5.3 The Normal Distribution

The most common and best known distribution is the normal or Gaussian distribution.

A probability distribution P is called a *normal distribution* if its density function is

$$f(x) = \frac{1}{\sqrt{2\pi}\sigma} e^{-\frac{1}{2}\left(\frac{x-m}{\sigma}\right)^2}, \text{ for } x \in \mathbb{R}, \qquad (5.8)$$

with parameters $m \in \mathbb{R}$ and $\sigma > 0$.

MAPLE ASSISTANCE 50

Define the density function of the normal distribution and then use it for making plots for a number of parameters m and σ.

```
>  f := proc(m,sigma)
   local w,x:
   w := 1/(sqrt(2*Pi)*sigma):
   unapply(w*exp((-((x-m)/sigma)^2/2)),x):
   end:
>  plot([f(20,1),f(20,2),f(20,3)],
   10..30,linestyle=[3,4,1],thickness = 2, color = BLACK);
```

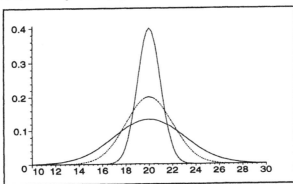

```
>  plot([f(15,3),f(20,3),f(25,3)],
   5..35,linestyle=[3,1,4], thickness = 2,color = BLACK);
```

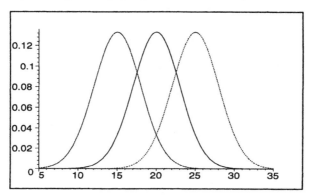

It is easy to see that $x = m$ is the maximum of the density function. Besides, one can easily check that the points $x = m - \sigma$ and $x = m + \sigma$ are the inflection points of this density.

We will use the following standard notation: $N(m, \sigma)$ denotes the normal distribution with parameters m and σ. The corresponding distribution function will be denoted by $\Phi_{m,\sigma}$. In particular, $\Phi_{0,1}$ denotes the distribution of the standardized normal distribution $N(0, 1)$ and is usually written as Φ. Thus,

$$\Phi(x) = \frac{1}{\sqrt{2\pi}} \int_{-\infty}^{x} e^{-\frac{1}{2}t^2} dt. \qquad (5.9)$$

We recall a simple geometric meaning of the distribution function.

MAPLE ASSISTANCE 51

We shade the area under the graph of the density function representing the value of $\Phi_{m,\sigma}(x)$ and define a procedure that can be used with any parameters m, σ and value of x.

```
>  with(plottools):
>  shade := proc(m,sigma,x):
>  seq(curve([[m + sigma*0.1*i,0], [m + sigma*0.1*i,
   f(m,sigma) (m + sigma*0.1*i)]], linestyle = 2), i =
   -30..30*(x-m)/(3*sigma))
   end:
```

```
Warning, 'i' in call to 'seq' is not local
>  f1 := plot(f(178,8),154..202,thickness=2,color=BLACK):
>  plots[display](f1,shade(178,8,186), xtickmarks=6,
   ytickmarks=2);
```

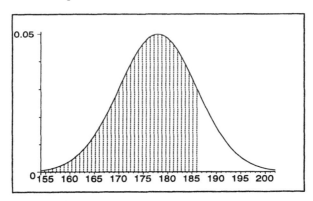

The values of $\Phi(x)$ are tabulated in nearly every textbook on probability and statistics, for example see page 139 in this book. Yet, a MAPLE user typically will not be interested in such a table.

MAPLE ASSISTANCE 52

We request the value of $\Phi(0.55)$.
> stats[statevalf,cdf,normald](0.55);
                          .7088403132

We request the value of $\Phi(-0.55)$
> stats[statevalf,cdf,normald](-0.55);
                          .2911596868

and the value of $\Phi^{-1}(0.975)$.
> stats[statevalf,icdf,normald](0.975);
                          1.959963985

We can obtain similar results for the normal distribution with any parameters. For example, the value of $\Phi_{20,3}(21.65)$ is
> stats[statevalf,cdf,normald[20,3]](21.65);
                          .7088403132

We note some useful properties of the distribution function Φ, which have an immediate geometric interpretation. Specifically we have

$$\Phi(0) = \frac{1}{2} \quad \text{and} \quad \Phi(x) = 1 - \Phi(-x) \text{ for every } x \in \mathbb{R}.$$

Changing the variable one can use Φ to compute $\Phi_{m,\sigma}$ for any parameters m and σ. Thus, we have

$$\Phi_{m,\sigma}(x) = \Phi\left(\frac{x - m}{\sigma}\right). \qquad (5.10)$$

The distribution function Φ if the standard normal distribution $N(0,1)$[1]

x	0.00	0.01	0.02	0.03	0.04	0.05	0.06	0.07	0.08	0.09
0.0	0.5000	0.5040	0.5080	0.5120	0.5160	0.5199	0.5239	0.5279	0.5319	0.5359
0.1	0.5398	0.5438	0.5478	0.5517	0.5557	0.5596	0.5636	0.5675	0.5714	0.5753
0.2	0.5793	0.5832	0.5871	0.5910	0.5948	0.5987	0.6026	0.6064	0.6103	0.6141
0.3	0.6179	0.6217	0.6255	0.6293	0.6331	0.6368	0.6406	0.6443	0.6480	0.6517
0.4	0.6554	0.6591	0.6628	0.6664	0.6700	0.6736	0.6772	0.6808	0.6844	0.6879
0.5	0.6915	0.6950	0.6985	0.7019	0.7054	0.7088	0.7123	0.7157	0.7190	0.7224
0.6	0.7257	0.7291	0.7324	0.7357	0.7389	0.7422	0.7454	0.7486	0.7517	0.7549
0.7	0.7580	0.7611	0.7642	0.7673	0.7704	0.7734	0.7764	0.7794	0.7823	0.7852
0.8	0.7881	0.7910	0.7939	0.7967	0.7995	0.8023	0.8051	0.8078	0.8106	0.8133
0.9	0.8159	0.8186	0.8212	0.8238	0.8264	0.8289	0.8315	0.8340	0.8365	0.8389
1.0	0.8413	0.8438	0.8461	0.8485	0.8508	0.8531	0.8554	0.8577	0.8599	0.8621
1.1	0.8643	0.8665	0.8686	0.8708	0.8729	0.8749	0.8770	0.8790	0.8810	0.8830
1.2	0.8849	0.8869	0.8888	0.8907	0.8925	0.8944	0.8962	0.8980	0.8997	0.9015
1.3	0.9032	0.9049	0.9066	0.9082	0.9099	0.9115	0.9131	0.9147	0.9162	0.9177
1.4	0.9192	0.9207	0.9222	0.9236	0.9251	0.9265	0.9279	0.9292	0.9306	0.9319
1.5	0.9332	0.9345	0.9357	0.9370	0.9382	0.9394	0.9406	0.9418	0.9429	0.9441
1.6	0.9452	0.9463	0.9474	0.9484	0.9495	0.9505	0.9515	0.9525	0.9535	0.9545
1.7	0.9554	0.9564	0.9573	0.9582	0.9591	0.9599	0.9608	0.9616	0.9625	0.9633
1.8	0.9641	0.9649	0.9656	0.9664	0.9671	0.9678	0.9686	0.9693	0.9699	0.9706
1.9	0.9713	0.9719	0.9726	0.9732	0.9738	0.9744	0.9750	0.9756	0.9761	0.9767
2.0	0.9772	0.9778	0.9783	0.9788	0.9793	0.9798	0.9803	0.9808	0.9812	0.9817
2.1	0.9821	0.9826	0.9830	0.9834	0.9838	0.9842	0.9846	0.9850	0.9854	0.9857
2.2	0.9861	0.9864	0.9868	0.9871	0.9875	0.9878	0.9881	0.9884	0.9887	0.9890
2.3	0.9893	0.9896	0.9898	0.9901	0.9904	0.9906	0.9909	0.9911	0.9913	0.9916
2.4	0.9918	0.9920	0.9922	0.9925	0.9927	0.9929	0.9931	0.9932	0.9934	0.9936
2.5	0.9938	0.9940	0.9941	0.9943	0.9945	0.9946	0.9948	0.9949	0.9951	0.9952
2.6	0.9953	0.9955	0.9956	0.9957	0.9959	0.9960	0.9961	0.9962	0.9963	0.9964
2.7	0.9965	0.9966	0.9967	0.9968	0.9969	0.9970	0.9971	0.9972	0.9973	0.9974
2.8	0.9974	0.9975	0.9976	0.9977	0.9977	0.9978	0.9979	0.9979	0.9980	0.9981
2.9	0.9981	0.9982	0.9982	0.9983	0.9984	0.9984	0.9985	0.9985	0.9986	0.9986
3.0	0.9987	0.9987	0.9987	0.9988	0.9988	0.9989	0.9989	0.9989	0.9990	0.9990

The parameters m and σ have a very important and simple probabilistic interpretation. Namely, we have the following theorem.

Theorem 5.3.1 *The expectation and variance of a random variable having normal distribution $N(m,\sigma)$ are*

$$\mathbb{E}(\xi) = m, \qquad \mathrm{Var}\,(\xi) = \sigma^2. \qquad (5.11)$$

Thus σ is the standard deviation.

---

[1] Tabulated here are values of $\Phi(x)$ for $0 \le x \le 3.09$.

We use MAPLE instead of a proof.

MAPLE ASSISTANCE 53

```
> assume(sigma > 0):
> E := int(x*f(m,sigma)(x),x=-infinity..infinity);
```
$$E := m$$
```
> Var := int(x^2*f(m,sigma)(x),x=-infinity..infinity) - E^2;
```
$$Var := \sigma^{\sim 2}$$

From Chebyshev's Inequality we already know that σ is a measure of concentration about the mean m. We can find more evidence for this in the case of the normal distribution. Compute the probabilities

$$r_k := P(m - k\sigma, m + k\sigma) \quad \text{for } k = 1, 2, 3, 4,$$

where P is the $N(m, \sigma)$ distribution, that is

$$r_k = \Phi_{m,\sigma}(m + k\sigma) - \Phi_{m,\sigma}(m - k\sigma) = \Phi(k) - \Phi(-k) = 2\Phi(k) - 1.$$

MAPLE ASSISTANCE 54

```
> seq(2*stats[statevalf,cdf,normald](i)-1,i = 1..4);
        .682689492, .954499736, .997300204, .999936658
```

As we have mentioned above, the normal (Gaussian) distribution is of fundamental importance. Many basic physical, biological, economic and other quantities share normal distributions with suitable parameters. The normal distribution is also the main key for statistical inference. An explanation of these phenomena is given by the Central Limit Theorem. Generally speaking, it says that the sum of a large number of independent random variables, under very natural assumptions, is approximately normally distributed.

5.4 Central Limit Theorem

The Laws of Large Numbers refer to the convergence of averages of independent random variables. They have a considerable amount of theoretical importance, but the practical value of these Laws seems to be fairly limited. Experience that the reader has already gained should suggest that all information about a random variable may be obtained from its distribution function. The Laws of Large Numbers say nothing about the distribution of the sum and say nothing interesting about the distribution of the average. The Central Limit Theorem fills in this gap.

Because of its importance we present the Central Limit Theorem in three, equivalent, versions. In all cases we consider the following situation.

Assumption.

Let (Ω, Σ, P) be a probability space and ξ_1, ξ_2, ξ_3, ... be independent random variables defined on Ω. Assume the ξ_i have a common distribution with finite expectation m and finite nonzero variance σ^2. (The latter means that the ξ_i are not constant.)
As usual define $S_n = \xi_1 + \ldots + \xi_n$.

Recall from Chapter 4 that the sum S_n and the average $\dfrac{S_n}{n}$ have expectations and standard deviations

$$\mathbb{E}(S_n) = nm, \qquad \sigma_{S_n} = \sigma\sqrt{n} \tag{5.12}$$

$$\mathbb{E}\left(\frac{S_n}{n}\right) = m, \qquad \sigma_{\frac{S_n}{n}} = \frac{\sigma}{\sqrt{n}} \tag{5.13}$$

We also consider the standardized sums

$$Z_n := \frac{S_n - \mathbb{E}(S_n)}{\sqrt{\mathrm{Var}\,(S_n)}} = \frac{S_n - nm}{\sigma\sqrt{n}},$$

for which it is easy to check that

$$\mathbb{E}(Z_n) = 0, \qquad \sigma_{Z_n} = 1. \tag{5.14}$$

Recall that F_X denotes the distribution function for any random variable X, i.e. $F_X(x) = P(X \leq x)$.

The following version is the celebrated Lindeberg-Lévy Theorem.

Theorem 5.4.1
$$\lim_{n \to \infty} F_{Z_n}(x) = \Phi(x),$$
for each $x \in \mathbb{R}$, where Φ is the distribution function of $N(0,1)$.

Proofs of this theorem are fairly long and tricky, so we do not give one here.

The Lindeberg-Lévy Theorem can be expressed in a more natural way without initial standarization. There are two versions, one for sums and the other for averages.

Theorem 5.4.2 *For each $x \in \mathbb{R}$*
$$\lim_{n \to \infty} \left(F_{S_n}(x) - \Phi_{nm, \sigma\sqrt{n}}(x) \right) = 0$$
and
$$\lim_{n \to \infty} \left(F_{\frac{S_n}{n}}(x) - \Phi_{m, \frac{\sigma}{\sqrt{n}}}(x) \right) = 0.$$

This theorem guarantees that for large n, typically $n \geq 30$ in practice, one can assume that the sum and the average of independent random variables have normal distributions. Moreover, Theorem 5.4.1 makes it possible to use the standardized normal distribution for a standardized sum, thus allowing the use of standard tables. In the computer era, we have straightforward access to the values of the normal distribution with any parameters, but the standardization is still useful, especially for statistical inference.

We emphasize that the claims in the Central Limit Theorems are valid under much more general assumptions than those above. For example, some kind of dependence is possible, the random variables may have different distributions, and more. The main idea is that each component of the sum S_n should be very small with respect to the sum itself.

We will give some empirical evidence for the Central Limit Theorem.

Example 5.4.1 We will simulate tossing a fair die and examine S_n, the sum of numbers received after n trials. Note that S_n has the form $S_n := \xi_1 + \ldots + \xi_n$, where each ξ_i denotes a number received in the i-th trial. Clearly, the ξ_i are independent and have a common discrete distribution concentrated at the points 1, 2, 3, 4, 5, 6 with equal probabilities $\frac{1}{6}$. One can easily determine the expectation and standard deviation

$$m = 3.5 \text{ and } \sigma = \frac{\sqrt{105}}{6}.$$

Assuming 1000 tosses ($n = 1000$) we know by the Central Limit Theorem that S_n has approximately the normal distribution $N(3500, \sqrt{1000}\frac{\sqrt{105}}{6})$. To verify this, we will simulate $n = 1000$ tosses of a die. Actually, we will repeat 400 such experiments to obtain 400 values of S_n. These values will give us an empirical distribution of S_n to be compared with the above theoretical one.

MAPLE ASSISTANCE 55

Fix k the number of series, and n, the number of tosses, in each series. Choose a die. The sums S_n will be collected in a list called *alist*.

```
>  k := 400: n := 1000: alist := NULL: die := rand(1..6):
```

Let us toss the die 400 000 times.

```
>  from 1 to k do
   S := 0:
   from 1 to n do
   S := S + die():
   od:
   alist := alist,S
   od:
```

We want to summarize the data obtained in classes to plot a histogram, using 18 classes each of length 20.

```
> data := stats[transform,tallyinto['outliers']]([alist],
    [seq(3320 + (i - 1)*20..3320 + i*20, i = 1..18)]);
```

$data :=$ [Weight(3560..3580, 29), Weight(3620..3640, 5),
Weight(3520..3540, 49), Weight(3540..3560, 36),
Weight(3380..3400, 10), Weight(3500..3520, 57),
Weight(3360..3380, 3), Weight(3480..3500, 59), 3340..3360,
Weight(3460..3480, 53), 3320..3340, Weight(3440..3460, 29),
Weight(3420..3440, 17), Weight(3600..3620, 9),
Weight(3400..3420, 22), 3660..3680, Weight(3580..3600, 16),
Weight(3640..3660, 2)]

Check how many items do not fall into any of these classes.

```
> outliers;
```
$$[3683]$$

Exactly one sum was too large. This does not matter.

We want to have probabilities instead of weights[2].

```
> data1 := stats[transform,
    scaleweight[1/nops([alist])]](data);
```

$data1 :=$ [Weight(3560..3580, $\frac{29}{400}$), Weight(3620..3640, $\frac{1}{80}$),
Weight(3520..3540, $\frac{49}{400}$), Weight(3540..3560, $\frac{9}{100}$),
Weight(3380..3400, $\frac{1}{40}$), Weight(3500..3520, $\frac{57}{400}$),
Weight(3360..3380, $\frac{3}{400}$), Weight(3480..3500, $\frac{59}{400}$),
Weight(3340..3360, $\frac{1}{400}$), Weight(3460..3480, $\frac{53}{400}$),
Weight(3320..3340, $\frac{1}{400}$), Weight(3440..3460, $\frac{29}{400}$),
Weight(3420..3440, $\frac{17}{400}$), Weight(3600..3620, $\frac{9}{400}$),
Weight(3400..3420, $\frac{11}{200}$), Weight(3660..3680, $\frac{1}{400}$),
Weight(3580..3600, $\frac{1}{25}$), Weight(3640..3660, $\frac{1}{200}$)]

---

[2] Actually, we should divide by 399 instead of 400 to avoid bias; see **E 4.10**. However, it does not make any practical difference here.

Plot the histogram:
```
> stats[statplots,histogram](data1);
```

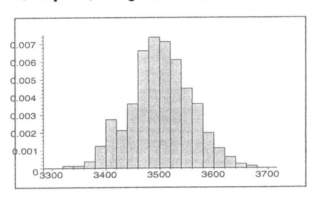

Save this histogram for further use.
```
> g1 := %:
```

Compute the theoretical values of m and σ.
```
> ed := add(i,i=1..6)/6: vd := add(i^2,i=1..6)/6 - ed^2:
> es := k*ed; vs := k*vd;
```
$$es := 3500$$
$$vs := \frac{8750}{3}$$

Make (but do not display yet) a plot of the density of the theoretical distribution, where the function f has already been defined.
```
> g2 := plot(f(es,sqrt(vs)),3320..3680, color=black):
```

Compute the empirical mean and standard deviation.
```
> ee := evalf(stats[describe,mean]([alist]));
```
$$ee := 3501.937500$$
```
> ve := evalf(stats[describe,standarddeviation]([alist]));
```
$$ve := 56.83892675$$

Make (but do not display yet) the plot of the density of the normal distribution with empirical distribution.
```
> g3 := plot(f(ee,ve),3320..3680, color=black,thickness=2):
```

Now display all three plots on the same picture.
```
> plots[display](g1,g2,g3);
```

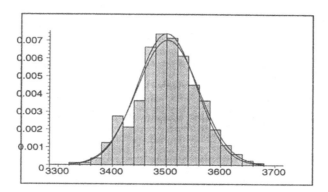

5.4.1 Examples

We consider another simple examples where the Central Limit Theorem is applicable.

Example 5.4.2 Compute the probability that after tossing a die 1000 times one receives more than 150 "6"s.

The number of "6"s we are interested in is the sum S_n, for $n = 1000$, of independent random variables ξ_i having zero-one distribution with $p = \frac{1}{6}$, the probability of success. We want to compute $P(S_n > 150)$, which is the same as $1 - F_{S_n}(150)$. By the Central Limit Theorem we may assume that $F_{S_n} \cong \Phi_{nm, \sigma\sqrt{n}}$ when $m = p$ and $\sigma = \sqrt{p(1-p)}$.

MAPLE ASSISTANCE 56

```
>  n := 1000: p := 1/6:
   mS := n*p: sigmaS := sqrt(n*p*(1-p)):
   1 - stats[statevalf,cdf,normald[mS,sigmaS]](150);
                       .9213503969
```

Example 5.4.3 *Example 4.4.1 revisited.*
Compute $P(|\xi - 500| \geq 50)$, where ξ is the sum S_n of $n = 1000$ independent random variables having zero-one distribution with $p = \frac{1}{2}$, the probability of success.

We want to compute $1 - (F_{S_n}(550) - F_{S_n}(450))$. As in the previous Example we have the following.

MAPLE ASSISTANCE 57

```
>  n := 1000: p := 1/2:
   mS := n*p: sigmaS := sqrt(n*p*(1-p)):
   stats[statevalf,cdf,normald[mS,sigmaS]](550) -
   stats[statevalf,cdf,normald[mS,sigmaS]](450);
                       .9984345978
```

> $1 - \%;$

.0015654022

Example 5.4.4 A political party ABC wants to know in advance the fraction of the population voting for ABC during the coming election. They commission an investigation in which people from a randomly chosen sample are asked. How large a sample should be used in order to estimate this fraction with accuracy $\varepsilon = 0.03$?

First of all note that the above question does not make sense literally. Even if all individuals from the sample say YES, it may happen that half of the population will say NO. However, the probability of such an event is very very small. We have to reformulate the problem. Instead, we ask: How large should a sample be so that the probability of the fraction obtained during the election will not differ from the fraction computed from the sample by more than $\varepsilon = 0.03$ is less than $1 - \alpha$, for some small α, e.g. $\alpha = 0.05$?

Let $p \in (0, 1)$ denote the true (still unknown) fraction of adherents of the party ABC. Let n be the size of a sample and let S_n denotes the number of adherents of ABC in the sample. Then $\dfrac{S_n}{n}$ is a good estimator of p. Clearly, S_n is the sum of n random variables, say ξ_i, having a zero-one distribution with p, the probability of success. If the sample was chosen randomly, we can assume that the ξ_i are independent. We want to find n such that

$$P\left(\left|\frac{S_n}{n} - p\right| \le \varepsilon\right) \ge 1 - \alpha.$$

Since, by the Central Limit Theorem, the average $\dfrac{S_n}{n}$ has approximately the normal distribution with parameters m and $\sqrt{\frac{p(1-p)}{n}}$, the above inequality is equivalent to:

$$P\left(|Z_n| \le \frac{\varepsilon\sqrt{n}}{\sqrt{p(1-p)}}\right) \ge 1 - \alpha.$$

Now

$$2\Phi\left(\frac{\varepsilon\sqrt{n}}{\sqrt{p(1-p)}}\right) - 1 \ge 1 - \alpha$$

can be solved for

$$n \ge \left(\frac{\Phi^{-1}\left(1 - \frac{\alpha}{2}\right)}{\varepsilon}\right)^2 (1-p)p.$$

Although p is not known, we have: $(1-p)p \le \frac{1}{4}$ for $0 \le p \le 1$, so an n satisfying

$$n \ge \left(\frac{\Phi^{-1}\left(1 - \frac{\alpha}{2}\right)}{\varepsilon}\right)^2 \frac{1}{4}$$

is a sufficient size for such a sample.

Moreover, in some cases we are sure that p is less than some $p_0 \leq \frac{1}{2}$. (The ABC leaders are aware of the fact that $p \leq 0.1$ and a reason for the investigation is to make sure that $p > 0.05$). In this case, since $p(1 - p) \leq p_0(1 - p_0)$, the estimation for n will be more accurate.

MAPLE ASSISTANCE 58

```
>   size := proc(alpha,p0,epsilon)
>   (stats[statevalf,icdf,normald](1-alpha/2)/
    epsilon)^2*p0*(1-p0)
    end:
>   size(0.05,0.1,0.03);
                        384.1458823
```

Inserting a MAPLE Spreadsheet, we can compare the sample size for various α, ε and p_0.

α	$p0$	ε	$stats_{statevalf,\,icdf,\,normald}(1 - \frac{1}{2}\alpha)^2 \, p0\,(1 - p0)$
			ε^2
.05	.5	.03	1067.071896
.1	.5	.03	751.5398485
.05	.1	.05	138.2925176
.1	.08	.03	221.2533313

5.5 Multidimensional Normal Distribution

We define an n-dimensional normal (Gaussian) distribution in a similar way as we did for the one dimensional case.

An n-dimensional probability distribution P is called a normal (Gaussian) distribution if its density function $f : \mathbb{R}^n \longrightarrow \mathbb{R}$ is

$$f(x) = \frac{1}{(2\pi)^{\frac{n}{2}}|\Sigma|^{\frac{1}{2}}} \exp\left(-\frac{1}{2}(x - \mu)^T \Sigma^{-1}(x - \mu)\right).$$

Here, the vector $\mu \in \mathbb{R}^n$ and the n-dimensional matrix Σ are given. We assume that Σ is a symmetric and positive definite (hence invertible) matrix, that is $\Sigma^T = \Sigma$ and $x^T \Sigma x > 0$ for any nonzero $x \in \mathbb{R}^n$. $|\Sigma|$ denotes the determinant of Σ and A^T denotes the transpose of A.

We will write $N_n(\mu, \Sigma)$ for the normal distribution with parameters μ and Σ. Let us note that the above definition is not the same as the definition on page 136 for the one dimensional case. In this case, $\mu = m$ and $\Sigma = \sigma^2$, which means that $N(m, \sigma)$ is the same as $N_1(\mu, \Sigma) = N_1(\mu, \sigma^2)$. Such, often misleading, notation is used for traditional reasons and the reader's convenience. In fact, most books and also MAPLE use such notation.

We have the following:

Theorem 5.5.1 Let $X = (\xi_1, \ldots, \xi_n)$ be a random vector $N_n(\mu, \Sigma)$ distributed.

Then each ξ_i is normal with

$$\mathbb{E}(\xi_i) = \mu_i \quad and \quad cov(\xi_i, \xi_j) = \Sigma_{ij} \quad for \ i, j = 1, \ldots, n. \tag{5.15}$$

In particular, $\mathrm{Var}\,(\xi_i) = \Sigma_{ii}$, $i = 1, \ldots, n$.

The distribution $N_n(0, I_n)$, where I_n denotes the identity matrix, is called the standard n-dimensional normal distribution. It is not hard to prove that if a random vector Z is $N_n(0, I_n)$ distributed, then the vector

$$X = AZ + \mu \tag{5.16}$$

is $N_n(\mu, \Sigma)$ distributed, with $\Sigma = A^T A = A^2$, for any vector $\mu \in \mathbb{R}^n$ and positively definite symmetric matrix A.

5.5.1 2-dimensional Normal Distribution

Now, consider a 2-dimensional $N_2(\mu, \Sigma)$ distributed vector $X = (\xi, \eta)$. In this case we have

$$\mu = \begin{bmatrix} m_\xi \\ m_\eta \end{bmatrix}, \quad \Sigma = \begin{bmatrix} \sigma_\xi^2 & \rho \, \sigma_\xi \sigma_\eta \\ \rho \, \sigma_\xi \sigma_\eta & \sigma_\eta^2 \end{bmatrix}. \tag{5.17}$$

and the formula for the density becomes

$$f(x, y) = \frac{1}{2\pi \, \sigma_\xi \, \sigma_\eta \, \sqrt{1 - \rho^2}} e^{-\frac{1}{2}\left(\frac{(x - m_\xi)^2}{\sigma_\xi^2} - 2 \frac{\rho \, (x - m_\xi)(y - m_\eta)}{\sigma_\xi \, \sigma_\eta} + \frac{(y - m_\eta)^2}{\sigma_\eta^2} \right)}. \tag{5.18}$$

The density plot and its contour plot help one to understand the meaning of the five parameters.

MAPLE ASSISTANCE 59

Define the density function of the two-dimensional normal distribution.

```
> coeffi := 1/(2*Pi*sigma1*sigma2*sqrt(1-rho^2));
```

$$coeffi := \frac{1}{2} \frac{1}{\pi \, \sigma 1 \, \sigma 2 \, \sqrt{1 - \rho^2}}$$

```
> f := (x,y) -> coeffi*exp(-1/2/(1-rho^2)*((x-m1)^2/sigma1^2
  -2*rho*(x-m1)*(y-m2)/sigma1/sigma2
  +(y-m2)^2/sigma2^2));
```

$$f := (x, y) \rightarrow coeffi \ e^{\left(-1/2 \frac{\frac{(x - m1)^2}{\sigma 1^2} - 2 \frac{\rho (x - m1)(y - m2)}{\sigma 1 \sigma 2} + \frac{(y - m2)^2}{\sigma 2^2}}{1 - \rho^2} \right)}$$

Consider some specific parameters.

```
>  m1 := 2: m2 := 3: sigma1 := 1: sigma2 := 2: rho := -0.7:
```

Plot the density and corresponding contourplot.

```
>  plot3d(f,-1..5,-5..11,style = wireframe,axes=FRAMED);
```

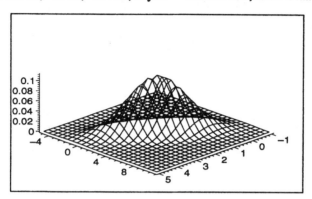

```
>  plots[contourplot](f,-1..5, -5..11,numpoints=2000,
   axes = BOXED);
```

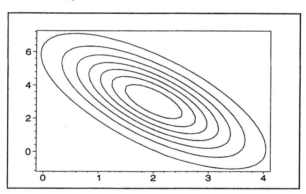

It is instructive to understand the meaning of ρ.

MAPLE ASSISTANCE 60

Define a list of parameters ρ, and use animation to display it.

```
>  K := [seq(-1+k*0.2,k=1..9)]:
>  anim := seq(plots[contourplot](f(x,y),
   x = -1..5, y = -5..11,
   numpoints=2000, tickmarks=[2,2],
   axes = BOXED),rho = K):
>  plots[display]([anim],insequence = true):
>  plots[display](%);
```

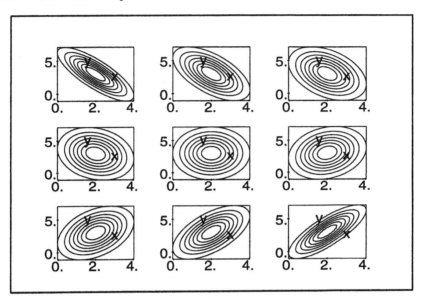

In the previous Chapter we showed that if the random variables are independent then their correlation coefficient is zero. Formula (5.18) indicates that in the case of a normal distribution the converse statement is also true. In fact, if $\rho = 0$ we see that $f(x, y) = f_\xi(x) f_\eta(y)$ which means independence of ξ and η.

One task which is often encountered is to select a sequence of N points that are distributed according to a specified normal distribution. Usually, the first step is to obtain N points having the standard normal distribution. The second step involves transforming these into points distributed according to the required distribution.

The first step can be performed in two ways. We can select two lists of N points each using the `random[normald]` procedure and concatenate them into a single list of points.

A more sophisticated approach is the *Box-Muller method*. It is based on the observation that if U_1 and U_2 are two independent $U(0, 1)$ uniformly distributed random variables then N_1 and N_2, defined by

$$N_1 = \sqrt{-2\ln(U_1)}\cos(2\pi U_2) \qquad (5.19)$$

$$N_2 = \sqrt{-2\ln(U_1)}\sin(2\pi U_2),$$

are two independent standard Gaussian random variables. This can be verified with a change of coordinates from Cartesian coordinates (N_1, N_2) to polar coordinates (r, θ) and then to $U_1 = \exp(-\frac{1}{2}r^2)$ and $U_2 = \theta/2\pi$.

We compare the above methods.

MAPLE ASSISTANCE 61

Call the `stats` package and define the Box-Muller transformation.

```
>  with(stats):
>  epi := evalf(Pi):
>  BM := (u1,u2) ->[sqrt(-2*ln(u1))*cos(2*epi*u2),
   sqrt(-2*ln(u1))*sin(2*epi*u2)];
```

$$BM := (u1,\ u2) \to [\sqrt{-2\ln(u1)}\cos(2\ epi\ u2),\ \sqrt{-2\ln(u1)}\sin(2\ epi\ u2)]$$

We will select 1000 points using both methods to compare the times used by each of them.

```
>  n := 1000:
```

Standard method.

```
>  at := time():
>  N1 := [random[normald](n)]: N2 := [random[normald](n)]:
>  Z := zip((x,y) ->[x,y],N1,N2):
>  time() - at;
```
$$4.952$$

Box-Muller method:

```
>  at := time():
>  U1 := [random[uniform](n)]: U2 := [random[uniform](n)]:
>  Z := transform[multiapply[BM]]([U1,U2]):
>  time() - at;
```
$$4.222$$

We can graphically verify the latter method. First we plot the drawn points.

```
>  plots[pointplot](Z);
```

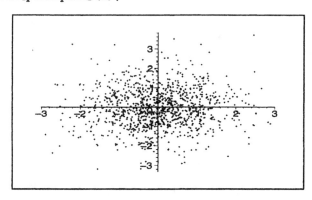

We can also make the histogram for all numbers obtained (2000 of them).

```
>  Y := map(op,Z):
>  statplots[histogram](Y,area = 1);
```

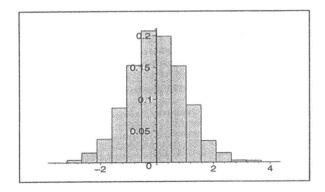

The second problem was to draw a sample from any two dimensional normal distribution with parameters m_1, m_2, σ_1, σ_2 and ρ. We can achieve it as follows. First, we need to find a positive definite symmetric matrix A of the form

$$A = \begin{bmatrix} a1 & b \\ b & a2 \end{bmatrix}.$$

Taking into account that

$$A^2 = \Sigma = \begin{bmatrix} \sigma_1^2 & \rho\sigma_1\sigma_2 \\ \rho\sigma_1\sigma_2 & \sigma_2^2 \end{bmatrix}$$

we have the equations $a1^2 + b^2 = \sigma_1^2$, $a2^2 + b^2 = \sigma_2^2$, $(a1 + a2)b = \rho\sigma_1\sigma_2$. If we can solve this system for $a1$, $a2$ and b we will be able to use formula (5.16) to convert data from the standard normal distribution. We present an example below.

MAPLE ASSISTANCE 62

Fix some parameters.
```
>  m1 := 20: m2 := 30: sigma1 := 2: sigma2 := 4: rho := 0.9:
```
Find $a1$, $a2$ and b.
```
>  rwn := {a1^2 + b^2 = sigma1^2, a2^2 + b^2 = sigma2^2,
   (a1 + a2)*b = rho*sigma1*sigma2};
      rwn := {a2² + b² = 16, (a1 + a2)b = 7.2, a1² + b² = 4}
>  fsolve(rwn,{a1,a2,b}):
>  assign(%);
      {b = -1.994945025, a2 = -3.467015193, a1 = -.1421068181}
```
Convert the data Z obtained from the Box-Muller method.
```
>  X := map(x ->[a1*x[1]+b*x[2]+m1,b*x[1]+a2*x[2]+m2],Z):
```
We can verify the results to some extent.

MAPLE ASSISTANCE 63
```
>  X1 := map(x->x[1],X): X2 := map(x->x[2],X):
>  describe[mean](X1); describe[mean](X2);
                  20.05747390
```

$$30.11488188$$
```
>  describe[standarddeviation](X1);
   describe[standarddeviation](X2);
```
$$2.037699118$$
$$4.093839654$$
```
>  describe[linearcorrelation](X1,X2);
```
$$.9068673862$$

We can also plot the data along with the density contourplot.
```
>  p1 := plots[pointplot](X,symbol = POINT):
>  p2 := plots[contourplot](f, 12..28, 12..48, contours =
   [0.0001,0.001,0.01,0.03], numpoints=2000, axes = BOXED):
>  plots[display](p1,p2);
```

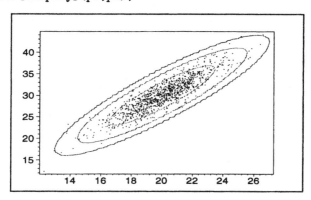

5.6 MAPLE Session

M 5.1 *Use formula (3.16) for the distribution of the sum of independent random variables taking non-negative integer values to find the distribution of the sum of independent random variables having Poisson distributions.*

Define a discrete convolution.
```
>  convd := proc(f,g);
   k -> sum(f(k-'i')*g('i'),'i'=0..infinity);
   end;
```
$$convd := \mathbf{proc}(f, g)\, k \rightarrow \text{sum}(f(k - \,'i') \times g('i'), \,'i' = 0..\infty)\, \mathbf{end}$$

Fix two Poisson distributions.
```
>  f := k -> a^k*exp(-a)/k!; g := k -> b^k*exp(-b)/k!;
```
$$f := k \rightarrow \frac{a^k\, e^{(-a)}}{k!}$$
$$g := k \rightarrow \frac{b^k\, e^{(-b)}}{k!}$$

The probability of the sum is thus:

```
>  convd(f,g)(k);
```

$$\frac{a^k\, e^{(-b-a)}\, (\frac{a+b}{a})^k}{k!}$$

```
>  normal(%);
```

$$\frac{a^k\, e^{(-b-a)}\, (\frac{a+b}{a})^k}{k!}$$

Despite its strange form, we can easily see that the above expression is also a Poisson distribution with parameter $a + b$.

M 5.2 *In Example 4.4.3 we saw, using Chebyshev's Inequality, that to have 95% certainty of getting 100 distinct elements out of a population of size 200 while drawing with replacement one needs to use at least 173 drawings. Can we improve on this result by applying the Central Limit Theorem?*

Actually, the answer should be no, as the random variable T defined in Example 4.3.5 and used in Example 4.4.3 does not satisfy the assumptions of the Central Limit Theorem. Nevertheless, we may still ask if T has a normal distribution. We will answer this by simulating T.

We call some needed packages.

```
>  with(stats):
>  with(stats[statplots]): with(stats[transform]):
>  with(plots):
```

Now, we perform *ntrials* = 500 experiments. Each time we are drawing without replacement elements from a population of size 200 until we get 100 distinct elements. The number of drawings needed are written down and form the sequence *data*.

```
>  get := rand(1..200):
>  ntrials := 500:
>  data := NULL:
>  from 1 to ntrials do
>  alist := NULL: n := 1: new := get():
>  while nops([alist]) < 100 do
   while member(new,[alist]) do
   new := get(): n := n+1 od;
   alist := alist,new:
   od:
   data := data,n:
   od:
```

We compute the mean and the standard deviation from the data. We also compute h, the optimal (according to certain criterion that we do not discuss here, see Silverman book [30]) width of the class-interval used in the following histogram plot.

```
> m := evalf(describe[mean]([data]));
  sigma := evalf(describe[standarddeviation]([data]));
  c := 3.486*sigma: h := evalf(c/ntrials^(1/3));
```

$$m := 138.9340000$$
$$\sigma := 7.614567880$$
$$h := 3.344382769$$

Let us make the histogram and compare it with the density of normal distributions $N(m, \sigma)$ with the parameters computed above. Note how many data points have not been used.

```
> k := floor(3*sigma/h):
  a := m - (k+0.5)*h: b := m + (k+0.5)*h:
  data1 := tallyinto['outliers']([data],
  [seq(a+i*h..a+(i+1)*h,i=0..2*k)]);
  data2 := scaleweight[1/nops([data])](data1):
  hist := histogram(data2,colour=cyan):
> pp := plot(stats[statevalf,pdf,
  normald[m,sigma]],a..b, color=black):
> display(hist,pp);
> outliers;
```

$$data1 := [\text{Weight}(133.9174259..137.2618086, 95),$$
$$\text{Weight}(120.5398948..123.8842775, 5),$$
$$\text{Weight}(123.8842775..127.2286603, 24),$$
$$\text{Weight}(117.1955120..120.5398948, 2),$$
$$\text{Weight}(130.5730431..133.9174259, 50),$$
$$\text{Weight}(153.9837225..157.3281052, 14),$$
$$\text{Weight}(140.6061914..143.9505742, 75),$$
$$\text{Weight}(143.9505742..147.2949569, 61),$$
$$\text{Weight}(150.6393397..153.9837225, 14),$$
$$\text{Weight}(157.3281052..160.6724880, 5),$$
$$\text{Weight}(127.2286603..130.5730431, 39),$$
$$\text{Weight}(137.2618086..140.6061914, 81),$$
$$\text{Weight}(147.2949569..150.6393397, 34)]$$

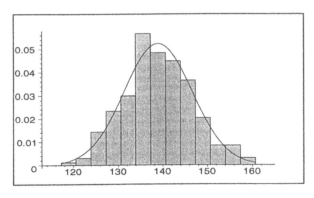

[165]

The above plot suggests that T may in fact have a normal (Gaussian) distribution. Moreover, in 6 out of 500 experiments the number of drawings exceeded 158, but was always at most 165.

Let us assume that T is normally distributed. With the mean and variance of T that we determined in Chapter 4 we can compute x such that $P(T < x) > 0.95$.

```
> x:=statevalf[icdf,normald[138.1306861,
  sqrt(60.37514711)]](0.95);
```
$$x := 150.9114366$$

This remarkably improves on the result of 173 drawings from Example 4.3.5.

M 5.3 *Instead of a histogram, we can plot a density function which approximates the density of the random variable T. The method used below gives the so called kernel estimator of the density.*

Define a kernel.
```
> K := stats[statevalf,pdf,normald]:
```

Sort our data to speed up the computation.
```
> sdata := sort([data]):
```

Define an optimal (in some sense, see for example [30]) number h.
```
> h := evalf(1.06*sigma*ntrials^(-1/5));
```
$$h := 2.374385747$$

Define the kernel estimator.
```
> ff := x ->1/(ntrials*h)*sum(K((x-sdata[i])/h),
  i = 1..ntrials);
```
$$ff := x \rightarrow \frac{\displaystyle\sum_{i=1}^{ntrials} K(\frac{x - sdata_i}{h})}{ntrials\ h}$$

Plot both the kernel estimator and the previous normal density function.

```
>   arange := 110..160:
>   wff := plot(ff,arange,thickness = 3):
>   plots[display]([pp,wff]);
```

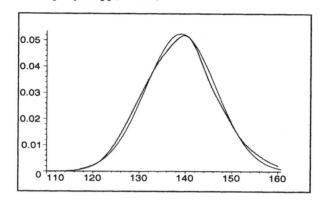

M 5.4 *Let R be a number of distinct elements drawn in a sequence of* 150 *drawings with replacement from a population of size* 200. *We simulate R to determine the shape of its distribution.*

```
>   ntrials := 500:
>   data := NULL:
>   from 1 to ntrials do
    alist := NULL:
>   from 1 to 150 do
    new := get():
    if not member(new,[alist]) then
    alist := alist,new fi:
    od:
    data := data,nops([alist]);
    od:
>   m := evalf(describe[mean]([data]));
    sigma := evalf(describe[standarddeviation]([data]));
```

$$m := 105.5580000$$
$$\sigma := 3.810857646$$

Again, we compare the kernel and the normal densities.

```
>   sdata := sort([data]):
>   h := evalf(1.06*sigma*ntrials^(-1/5));
```

$$h := 1.165559881$$

```
> ff := x ->1/(ntrials*h)*sum(K((x-sdata[i])/h),
  i = 1..ntrials):
> arange := 90..120:
> wff := plot(ff,arange,thickness = 3):
> pp := plot(stats[statevalf,pdf,normald[m,sigma]], arange,
  color = black):
> plots[display]([pp,wff]);
```

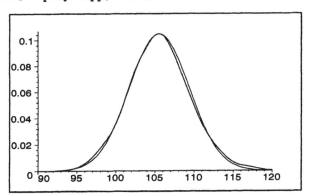

5.7 Exercises

E 5.1 Establish a situation in which the sum of two independent random variables with a binomial distribution also has a binomial distribution.

E 5.2 Prove that the sum of independent normally distributed random variables is also normally distributed.

E 5.3 Find $P(X > \mathbb{E}(X) + 3\sqrt{\text{Var}(X)})$, where X is a random variable having (a) uniform distribution on the interval $(0, 20)$, (b) normal distribution $N(10, 3)$, (c) exponential distribution with parameter $\lambda = 0.1$.

E 5.4 Let S denote the number of successes in 200 Bernoulli trials with 0.01 as the probability of success in each trial. Using Poisson and normal distributions compute (a) $P(S > 0)$, (b) $P(S < 0)$, (c) $P(S < 4)$, (d) $P(S = 2)$. Compare the probabilities. Which approximation is better?

E 5.5 Find the distribution of $\min(\xi, \eta)$, when ξ and η are independent random variables having exponential distributions.

E 5.6 A random variable ξ has normal distribution $N(m, \sigma)$. Find the distribution of e^{ξ}.

E 5.7 How large should a sample be so that it contains at least 50 females, assuming an equal probability for selecting females and males in the population under consideration. Use a confidence level of 95%.

E 5.8 How many raisins in a roll should a baker plan for to be 95% sure that a randomly chosen roll contains at least 5 raisins?

E 5.9 Compute the probability that the roots of the equation $x^2 + px + q$ are real numbers, if one knows p and q to be independent random variables that are uniformly distributed on the interval $[-1, 1]$.

E 5.10 Prove that the random variable $\dfrac{\xi}{\xi + \eta}$ is uniformly distributed on the interval $[0, 1]$, provided ξ and η are independent and have a common exponential distribution.

E 5.11 Let ξ and η be independent and $N(0, 1)$ distributed. Are $\xi + \eta$, $\xi - \eta$ independent?

E 5.12 Let $\xi_1, \xi_2, \xi_3, \ldots$ be independent random variables that are uniformly distributed on the interval $(0, 1)$. (a) Find the density of the sums and S_2, S_3, S_4 and plot them. (b) Compute $P(S_{100} \geq 70)$.

E 5.13 Solve Exercise 4.8 for $N = 100$ and $N = 1000$.

E 5.14 1000 throws of a fair coin have been performed. Compute the probability that the number of heads is in the interval (a) $(490, 510)$, (b) $(450, 550)$, (c) $(500, 600)$. Try to predict these probabilities before doing the computation.

E 5.15 How many throws of a fair die should be made to be 95% sure that a "6" will appear in at least 15 % of tosses?

6 Numerical Simulations and Statistical Inference

In this Chapter we present methods that are used for generating random or more precisely pseudo-random numbers. Then, we discuss briefly concepts and routines that are basic for statistical inference. Some of them are useful to test the quality of pseudo-random numbers and in statistical modeling.

6.1 Pseudo-Random Number Generation

In previous chapters we asked MAPLE to generate random numbers having certain properties, or more formally having a specific distribution. MAPLE like most computer algebra packages has a built-in random number generator. We will discuss this concept in more detail as the problem of random number generation is also very important in modeling real systems. The numerical simulation of a mathematical model of a complicated probabilistic system often provides information about the behaviour of the model that cannot be obtained directly or easily by other means. Numerical values of each of the random variables must be provided for a test run on the model, and then the outputs of many test runs are analyzed statistically. This procedure requires the generation of large quantities of random numbers with specified statistical properties.

Originally such numbers were taken directly from actual random variables, generated, for example, mechanically by tossing a die or electronically by the noisy output of a valve, and often listed in random number tables. This proved impractical for large scale simulations and the numbers were not always statistically reliable. In addition, a particular sequence of random numbers could not always be reproduced, an important feature for comparative studies, and so had to be stored. The advent of electronic computers lead to the development of simple deterministic algorithms to generate sequences of random variables quickly and reproducably. Such numbers are consequently not truly random, but with sufficient care they can be made to resemble random numbers in most properties, in which case they are called *pseudo-random numbers*.

These days most digital computers include a *linear congruential pseudo-random number generator*. These have the recursive form

$$X_{n+1} = aX_n + b \pmod{p} \tag{6.1}$$

where a and p are positive integers and b is a non–negative integer. For an integer initial value or *seed* X_0, algorithm (3.1) generates a sequence taking integer values from 0 to $p - 1$, i.e. the remainders when the $aX_n + b$ are divided by p. When the coefficients a, b and p are chosen appropriately the numbers

$$U_n = X_n/p \tag{6.2}$$

seem to be uniformly distributed on the unit interval $[0, 1]$. Since only finitely many different numbers occur, the modulus p should be chosen as large as possible, and perhaps also as a power of 2 to take advantage of the binary arithmetic used in computers. To prevent cycling with a period less than p the multiplier a should also be taken relatively prime to p. Typically b is chosen equal to zero, the resulting generator then being called a *multiplicative generator*. A much used example was the *RANDU* generator of the older *IBM* Scientific Subroutine Package with multiplier $a = 65,539 = 2^{16} + 3$ and modulus $p = 2^{31}$; the *IBM* System 360 Uniform Random Number Generator uses the multiplier $a = 16,807 = 7^5$ and modulus $p = 2^{31} - 1$, which is a prime number.

MAPLE uses a multiplicative generator working with larger a and prime p, see below. Actually, there are three standard options of obtaining pseudo-random numbers within MAPLE:

1. Obtaining a non–negative integer from the given range.
2. Obtaining a 12-digital integer.
3. Obtaining a sequence of the required length of numbers from one of the built-in distributions. Here MAPLE offers some extra options, see MAPLE Help for details.

We illustrate the above with examples.

MAPLE ASSISTANCE 64

Define the procedure (use a semicolon ; to see its code):
> die := rand(1..6);

> die := **proc**()
>   **local** t;
>   **global** $\_seed$;
> $\_seed$:= irem($427419669081 \times \_seed$, 999999999989) ;
> $t := \_seed$;
>   irem($t, 6$) + 1
> **end**

In MAPLE, irem means the remainder. For example:
> irem(23,5);

3

Now, we can call the above procedure as many times as we want.

```
>  seq(die(),i = 1..10);
```
$$1, 3, 1, 5, 5, 5, 4, 6, 2, 2$$

We may call the procedure rand with no parameters.

```
>  rand();
```
$$389023779969$$

We can see the code:

```
>  showstat(rand);
```

```
rand  :=  proc(r)

local a, p, s;

global _seed;

   1    p   :=  999999999989;

   2    a   :=  427419669081;

   3    s   :=  1000000000000;

   4    if not assigned(_seed) then

   5       _seed  :=  1

       fi;

   6    if nargs = 0 then

   7       _seed  :=  irem(a*_seed,p)

       else

   8       subs({_MODULUS = p, _MULTIPLIER = a, _SHIFT = s,
'rand/generator'(r,p)},proc ()

local t;

global _seed;

_seed :=  irem(_MULTIPLIER*_seed,_MODULUS);

t   :=  _seed;

to _CONCATS do

_seed   :=  irem(_MULTIPLIER*_seed,_MODULUS);

t   :=  _SHIFT*t+_seed

od;

irem(t,_DIVISOR)+_OFFSET

end)
```

```
        fi

  end
```

We can check the current value of the variable _seed.

```
  > _seed;
```
$$389023779969$$

Sometimes it is convenient to fix the value of the _seed. For example, we may want to have the same sequence of random numbers to try different procedures on them. Alternatively, we may want different random sequences each time. The latter, for example, can be done as follows.

```
  > _seed := 1 + floor(1000*time());
```
$$\_seed := 1459$$

We can also generate, pseudo-random numbers with a specific distribution.

```
  > with(stats):
  > random[poisson[6]](15);
```
$$6.0,\ 6.0,\ 5.0,\ 10.0,\ 8.0,\ 6.0,\ 8.0,\ 8.0,\ 7.0,\ 8.0,\ 3.0,\ 5.0,\ 3.0,\ 3.0,\ 2.0$$

· · ·

Note, that the problem of generating pseudo-random numbers with the given distribution can be reduced to the problem of generating numbers with the uniform distribution on the interval $(0, 1)$. In fact, assume that the random variable ξ is uniformly distributed on the interval $(0, 1)$, i.e.

$$P(\xi \leq x) = \begin{cases} 0, & x < 0 \\ x, & 0 \leq x < 1 \\ 1, & 1 \leq x \end{cases}$$

Let F be a specific distribution function and assume for simplicity that F is strictly increasing. Define a new random variable η by $\eta = F^{-1}(\xi)$. For any x we then have

$$P(\eta \leq x) = P(F^{-1}(\xi) \leq x) = P(\xi \leq F(x)) = F(x).$$

Hence, F is the distribution function of η. In other words, if numbers x_i are uniformly distributed on $(0, 1)$, then the numbers $F^{-1}(x_i)$ are F distributed. As an example we will generate 200 numbers from exponential distributions with parameter $\lambda = 0.25$. Recall that its distribution function F is zero for negative x and $F(x) = 1 - e^{-\lambda x}$ for $x \geq 0$.

MAPLE ASSISTANCE 65

```
  > F := x ->1 - exp(-0.25*x);
```
$$F := x \to 1 - e^{(-.25\,x)}$$

```
> invF := y ->solve(F(x) = y,x);
```
$$invF := y \to \text{solve}(F(x) = y, \, x)$$

For example,
```
> invF(0.3);
```
$$1.426699776$$

Now, generate 200 uniformly distributed numbers from the interval $(0, 1)$.
```
> xi := [random[uniform](200)]:
```

And 200 numbers from the exponential distribution.
```
> eta := map(invF,xi):
```

Examine the histogram.
```
> statplots[histogram](eta, area = 1);
```

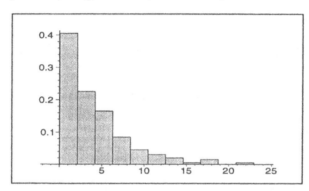

We see that the above plot corresponds to the density plot of the exponential distribution.

The Box-Muller method for generating normally distributed random numbers also uses uniformly distributed numbers, see page 150 in the previous chapter.

6.2 Basic Statistical Tests

In this section we discuss basic concepts of hypothesis testing. We confine ourselves to those aspects only that we will need to present tests concerning pseudo-random numbers.

Imagine we are given a sample $x_1, \ldots, x_n$ such that each x_i may be thought as a realisation of some random vector X_i. Assume that the vectors X_i have the same probability distribution, say P. On the basis of the sample we want to check some hypothesis, say H_0, concerning the common distribution P. For example, we may want to check the hypothesis that the average age of the university student is 22 on the basis of 100 student questionnaires. We proceed according to the following methodology. First of all we have to find a function, say, T, called a *statistic,* such that the quantity $T(x_1, \ldots, x_n)$ gives us some reasonable impression whether the sample confirms or does not confirm our hypothesis H_0. Under some general assumptions the composition $T \circ (X_1, \ldots, X_n)$, traditionally denoted by $T(X_1, \ldots, X_n)$, is a random vector and therefore we can talk about the distribution of the statistic T, understood as the distribution of $T(X_1, \ldots, X_n)$, under the assumption that the distribution P satisfies our hypothesis H_0. Fix some small positive α, called a *level of significance* and choose a set K according to the following rules:

(1) $T(X_1, \ldots, X_n) \in K$ strongly contradicts the hypothesis H_0,

(2) $P(K) = \alpha$ (sometimes $P(K) \cong \alpha$).

Such a K is called a *critical set.* Now, there are two cases:

(A) $T(x_1, \ldots, x_n) \in K$. Then we reject the hypothesis H_0.

(B) $T(x_1, \ldots, x_n) \notin K$. Then we do not have sufficient reason to reject the hypothesis H_0.

In many cases the choice of a critical set K is natural and simple. Generally it depends on accepting some alternative hypothesis to H_0, say H_1. We do not discuss this problem in the book.

Example 6.2.1 We want to check the hypothesis that the average age of the university student is 22. From $n = 100$ randomly chosen student questionnaires we have 100 numbers,

$$19, 19, 27, \ldots, 24, 22, 23.$$

According to the above scenario we have to establish some statistic T. Note, that in this situation we can choose among a few statistics, for example

$$T(x_1, \ldots, x_n) = \bar{x} = \frac{x_1 + \ldots + x_n}{n},$$

$$T(x_1, \ldots, x_n) = \frac{\max(x_1, \ldots, x_n) + \min(x_1, \ldots, x_n)}{2}$$

$$T(x_1, \ldots, x_n) = \mathrm{median}(x_1, \ldots, x_n),$$

where the *median* of the sample is its middle element, for n odd, and the average of its two middle elements, for n even, after sorting the sample. In all these cases we feel that the values of $T(x_1, \ldots, x_n)$ equal, for example, to 21.89 would confirm H_0 while $T(x_1, \ldots, x_n) = 25.23$ would strongly contradict it. There exists a theory which can help us to choose the most appropriate statistics, but we do not discuss the details here. In our particular case we will choose the first statistic as we actually know its distribution. Namely, if the sample comes from independent observations, we may assume that the random variables X_i are independent. Assume that they have the same distribution as the finite mean m and finite variation σ^2. From the Central Limit Theorem we then have that $T(X_1, \ldots, X_n)$ is normally distributed with mean m and variance $\frac{\sigma^2}{n}$. Moreover, we assume that $m = 22$. In this case we may also assume, by the Law of Large Numbers, see E 4.5.1, that $\sigma^2 = \frac{1}{n} \sum_{i=1}^{n} (m - x_i)^2$. Thus, we have fully determined the distribution of $T(X_1, \ldots, X_n)$ to be $\Phi_{m, \frac{\sigma}{\sqrt{n}}}$. Fix the level of significance, $\alpha = 0.05$. The critical set K may be defined in a few different ways. A natural choice in our situation is to have K symmetric about the mean, i.e.

$$K = (-\infty, m - \varepsilon) \cup (m + \varepsilon, \infty)$$

with ε such that $P(K) = \alpha$. Since we know the distribution, we can determine ε, and hence K itself. In fact, we have

$$\alpha = P(K) = \Phi_{m, \frac{\sigma}{\sqrt{n}}}(m - \varepsilon) + (1 - \Phi_{m, \frac{\sigma}{\sqrt{n}}}(m + \varepsilon)) = 2 - 2\Phi\left(\frac{\varepsilon\sqrt{n}}{\sigma}\right).$$

Hence $\varepsilon = \frac{\sigma}{\sqrt{n}} \Phi^{-1}\left(1 - \frac{\alpha}{2}\right)$.

Imagine that computed quantities from the sample are $\bar{x} = 22.43$ and $\sigma = 2.37$. Using MAPLE we find that $\varepsilon = 0.4645114644$. Thus we have that $\bar{x} \notin K$, hence we do not reject the hypothesis H_0.

Let us note that in a hypothesis testing routine we can never formally accept the hypothesis! On the other hand, in many practical situations we have to be more definite and say how strongly we believe that the hypothesis H_0 is true. There is not a good general solution to this problem. A partial solution is to perform another test using different statistics to obtain more evidence. Below, we discuss the concept of the p–value, which can also help in approaching a final decision.

6.2.1 p–value

Assume that we have chosen some statistic T to test the hypothesis and have already decided on the particular form of the critical set K_α for the level of significance α. Note that an increase of α results in an increase of the critical set K_α. The smallest α such that $T(x_1, \ldots, x_n) \in K_\alpha$ is called the *p-value*. In other words, the p–value is the smallest level of significance for which we would reject the hypothesis given the information from the sample. If the level of significance α is less than p-value then we would not reject the hypothesis. We can also think of the p-value as the probability of obtaining a value of the statistic as extreme as, or more extreme than, the actual value obtained when the hypothesis H_0 is true.

The practical meaning of the p–value is that the greater the p–value the better the chance of H_0 being true.

Example 6.2.2 We can compute the p–value in the previous Example. The observed value of the mean, $\bar{x} = 22.43$, and the assumed form of the critical set imply that we put $\varepsilon = \bar{x} - m = 0.43$. Thus we have

$$p\text{-value} = P(K) = \Phi_{m, \frac{\sigma}{\sqrt{n}}}(m - \varepsilon) + (1 - \Phi_{m, \frac{\sigma}{\sqrt{n}}}(m + \varepsilon)) = 2 - 2\Phi\left(\frac{\varepsilon\sqrt{n}}{\sigma}\right).$$

Using MAPLE we then have a p-value $= 0.069624475$.

The concept of p-value has been developed in recent years since fast computers have been available to handle the necessary calculations. In the era of statistical tables researchers in most cases were not able to compute the p-values.

6.3 The Runs Test

We will establish a test that the sample $x_1, \ldots, x_n$ is *simple*, i.e. it represents the sequence of independent random variables $X_1, \ldots, X_n$.

Let us imagine that we have the following sequence of numbers as a sample:

$$x = \{0.01, 0.02, 0.03, \ldots, 0.99\}.$$

We can immediately check that these numbers come from the $U(0, 1)$ distribution. The histogram, the mean, the variance and many other statistics show perfect compatibility with this distribution. On the other hand, we feel that it is not a simple sample at all in the above defined sense. The reason is that the numbers were chosen by using a very simple deterministic procedure. Since the pseudo-random numbers are also produced by a deterministic procedure, we should have a method to decide whether or not they can be still considered as random numbers. One such method is the *runs test*. We explain briefly its idea and present an example.

Consider first a sample of zeros and ones, say

$$1, 1, 1, 0, 1, 0, 1, 1, 1, 0, 0, 1, 0, 0, 0, 0, 0, 1, 1, 1.$$

A *run* is a sequence of zeros or ones in the sample that are preceded and followed by the opposite element or no element at all. In our sequence we thus have 9 runs. Further discussion is possible because of the intuitively obvious observation that if a sample contains too many or two few runs it is not a simple sample. For example, the above sample seems to be a simple sample contrary to the sample

$$1, 1, 1, 1, 1, 1, 1, 1, 1, 0, 0, 0, 0, 0, 0, 0, 0, 0, 1, 1.$$

and the sample

$$1, 0, 1, 0, 1, 0, 1, 0, 1, 0, 1, 0, 1, 0, 1, 0, 1, 0, 1, 1.$$

which have 3 and 19 runs respectively.

This is formally justified as follows. If the sequence containing n_1 zeros and n_2 ones is randomly chosen i.e. the sample is simple, then it is known that the number of runs R in the sample is a random variable with the distribution

$$P(R = k) = \begin{cases} 2 \dfrac{\dbinom{n_1 - 1}{k/2 - 1}\dbinom{n_2 - 1}{k/2 - 1}}{\dbinom{n_1 + n_2}{n_1}} & \text{for even } k \\[2em] \dfrac{\dbinom{n_1 - 1}{(k-1)/2}\dbinom{n_2 - 1}{(k-1)/2 - 1} + \dbinom{n_2 - 1}{(k-1)/2}\dbinom{n_1 - 1}{(k-1)/2 - 1}}{\dbinom{n_1 + n_2}{n_1}} \\ \text{for odd } k, \end{cases}$$

(6.3)

$k = 2, 3, \ldots, 2\min(n_1, n_2) + 1$. We will use R as the statistic to check the hypothesis.

For example, we make a plot of the above distribution for $n_1 = 11$ and $n_2 = 9$.

MAPLE ASSISTANCE 66

First we define a procedure.

```
>  runsdistr := proc(n1,n2,k)
   if type(k,even) then
   2*binomial(n1-1,k/2-1)*binomial(n2-1,k/2-1)/
   binomial(n1+n2,n1)
   else
   (binomial(n1-1,(k-1)/2)*binomial(n2-1,(k-1)/2-1) +
   binomial(n2-1,(k-1)/2)*binomial(n1-1,(k-1)/2-1) )/
   binomial(n1+n2,n1)
   fi:
   evalf(%)
   end:
```

And then plot the graph.

```
>  plots[pointplot](zip((x,y) ->[x,y],[$2..19],
   [seq(runsdistr(11,9,i),i=2..19)]),symbol = circle);
```

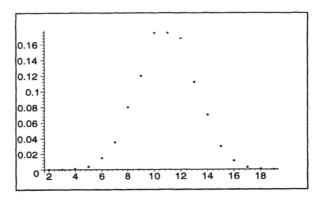

Assume now that we are given a sample, say $x_1, \ldots, x_n$, and we want to decide if it was randomly chosen. For simplicity we assume that the sample is drawn from a continuous distribution, which allows us to assume that all the elements of the sample are distinct from each other and are not equal to the median. Note, for a discrete distribution the runs test also works. The procedure is as follows:

1. Compute the median, i.e. the number μ such that half of the sample is below and the other half is above μ.
2. Transform the sample into a binary sample of zeros and ones putting for each $i = 1, \ldots, n$, 0 if $x_i < \mu$ and 1 otherwise.
3. Compute n_1, n_2, the numbers of ones and zeros, and k, the number of runs in the binary sample. Hence k is the value of the statistic R computed on the sample.
4. Using a standard statistical approach decide if the sample is simple, comparing k with the distribution of runs for determined n_1 and n_2. More specifically, we fix α, the level of significance, and find the critical set K covering extreme values such that $P(K) \cong \alpha$. Now, if $k \in K$ then we reject the hypothesis that the sample is simple. If $k \notin K$ we have no reason to reject the hypothesis.

We explain the procedure with the following examples.

Example 6.3.1 Draw a sample of size 20 from an exponential distribution and decide whether or not the sample may be considered as a simple sample.

MAPLE ASSISTANCE 67

First, we establish some useful procedures.

```
>  with(stats):
>  binX := proc(X::list)
   local Y, i, m:
   m := describe[mean](X); Y := NULL:
   for i from 1 to nops(X) do
   if X[i] < m then Y := Y,0 else Y := Y,1 fi
```

```
      od:
      [Y]:
      end:
>   runs := proc(X::list)
      local s, i:
      s := 1: i := 1:
>   while i < nops(X) do
      if X[i+1] <> X[i] then s := s + 1 fi:
      i := i+1:
      od:
>   s:
      end:
>   F := (n1,n2,r) -> add(runsdistr(n1,n2,i), i = 1..r):
```

Warning, 'i' in call to 'add' is not local

Obtain a sample.

```
>   X := [random[exponential[0.05]](20)];
```

$$X := [11.15204475, 7.745943774, 8.420706150, 12.85882300,$$
$$16.34967523, 27.46786522, .6517494560, 25.67288708,$$
$$18.54226236, 27.37537962, 6.017020142, 7.423461782,$$
$$31.90867762, .7991472476, 1.851750370, 64.62850182,$$
$$33.52442788, 12.09346153, 20.65825482, 50.67142316]$$

Transform the sample into a binary sample.

```
>   bX := binX(X);
```

$$bX := [0, 0, 0, 0, 0, 1, 0, 1, 0, 1, 0, 0, 1, 0, 0, 1, 1, 0, 1, 1]$$

Compute the numbers of ones and zeros.

```
>   n1 := add(bX[i], i = 1..nops(bX)); n2 := nops(bX) - n1;
```

$$n1 := 8$$
$$n2 := 12$$

Compute the number of runs.

```
>   k := runs(bX);
```

$$k := 12$$

Compare graphically the distribution of runs with k.

```
>   nn := 2*min(n1,n2) + 1;
```

$$nn := 17$$

```
>  prplot := plots[pointplot](zip((x,y) ->[x,y],[$2..nn],
   [seq(runsdistr(n1,n2,i),i=2..nn)]),symbol = circle):
   pline := plots[pointplot]([[k,0],[k,runsdistr(n1,n2,k)]],
   connect = true,linestyle = 3):
>  plots[display](prplot,pline);
```

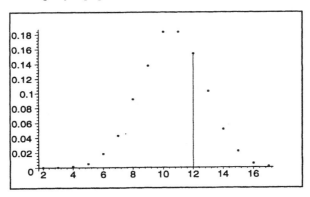

Let the level of significance be $\alpha = 0.1$. Since too small a number of runs and too large a number of runs contradict the hypothesis that the sample is random we choose the critical set K of the form $[2, \ldots, k_1] \cup [k_2, \ldots nn]$ such that

$$P([2, \ldots, k_1]) \cong P([k_2, \ldots, 2\min(n_1, n_2) + 1]) \cong \frac{\alpha}{2}.$$

```
>  alpha := 0.1:
```

We plot the graph of the distribution together with two lines such that points outside the strip between them correspond to the critical set.

```
>  cdfplot := plots[pointplot](zip((x,y) ->[x,y],[$2..nn],
   [seq(F(n1,n2,i),i=2..nn)]),symbol = circle):
   alphalines := plot([alpha/2,1-alpha/2], 2..nn,
   thickness = 3, color = BLACK):
>  plots[display]([cdfplot,alphalines]);
```

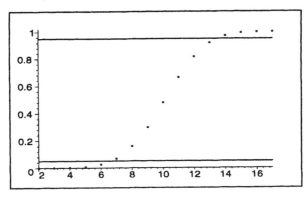

Inspect the plot.

```
>   F(n1,n2,6), 1- F(n1,n2,13);
                    .02460903390, .0798999762
```

This means that the critical set is $[2, \ldots 6] \cup [14, \ldots, 17]$. Thus, there is no reason to reject the hypothesis.

If the size of a sample is large and n_1, n_2 are not too small, then the distribution of runs is approximately the same as the normal distribution with appropriate mean m and standard deviation σ, namely,

$$m = \frac{2n_1 n_2}{n_1 + n_2} + 1, \quad \sigma = \sqrt{\frac{(m-1)(m-2)}{n_1 + n_2 - 1}}.$$

Even in the above situation when n is not large, this approximation is not bad. The advantage is that we can find the critical set K in a simpler way.

MAPLE ASSISTANCE 68

```
>   mR := 2*n1*n2/(n1+n2) + 1;
```
$$mR := \frac{53}{5}$$
```
>   stdR := sqrt((mR-1)*(mR-2)/(n1+n2-1));
```
$$stdR := \frac{4}{95} \sqrt{2451}$$
```
>   nplot := plot(statevalf[pdf,normald[mR,stdR]],
    mR - 3*stdR..mR + 3*stdR):
>   plots[display](prplot,nplot);
```

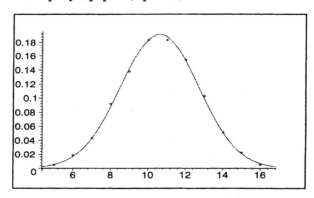

Now we can specify again a critical set K.

```
>   alpha := 0.1:
>   statevalf[icdf,normald[mR,stdR]](alpha/2),
    statevalf[icdf,normald[mR,stdR]](1 - alpha/2);
                    7.171254123, 14.02874588
```

We can now examine larger samples and lesser α.

MAPLE ASSISTANCE 69

```
>  X := [random[uniform](2000)]:
>  bX := binX(X):
>  n1 := add(bX[i], i = 1..nops(bX)); n2 := nops(bX) - n1;
```
$$n1 := 1007$$
$$n2 := 993$$

```
>  k := runs(bX);
```
$$k := 962$$

```
>  nn := 2*min(n1,n2) + 1:
>  mR := 2*n1*n2/(n1+n2) + 1:
>  stdR := sqrt((mR-1)*(mR-2)/(n1+n2-1)):
>  alpha := 0.01:
>  statevalf[icdf,normald[mR,stdR]](alpha/2),
   statevalf[icdf,normald[mR,stdR]](1 - alpha/2);
```
$$943.3709371, 1058.531063$$

Thus, the number of runs k does not belong to the critical set and we do not reject the hypothesis.

6.4 Goodness of Fit Tests

Another fundamental question when we work with pseudo-random numbers is whether or not a sample obtained represents the required distribution. Actually, this problem is even more important if we have to examine a sample resulting from real observations. In particular, we will discuss the following problem. Let $x_1, \ldots, x_n$ be a simple sample. We want to check the hypothesis that the sample comes from a given distribution F. We describe one of the goodness of fit tests that are used to manage this problem. It is the so called χ^2-goodness of fit test or χ^2-Pearson test. We outline the idea of the test.

Partition the sample into R mutually disjoint cells, $C_1, \ldots, C_R$ having $w_1, \ldots, w_R$ elements, respectively. Clearly $\sum_{i=1}^{R} w_i = n$. Then we calculate probabilities $P(C_1), \ldots, P(C_R)$ of the cells using the distribution F. We believe that if the sample comes from the desired distribution, then the frequencies $w_1, \ldots, w_R$ cannot be much different from the theoretical frequencies $nP(C_1), \ldots, nP(C_R)$. In particular, the quantity

$$h = \sum_{i=i}^{R} \frac{(w_i - nP(C_i))^2}{nP(C_i)}$$

should be small. It appears that under some mild assumptions the above statistic has the so called χ^2-distribution with a parameter known as the *number of degrees of freedom* that is equal to $R - 1$. In practice, the assumptions indicate that the size of the sample n should be greater than 30, the cells should contain similar numbers of elements and not less than 5 in each. For a given level of significance α the critical set K is the interval

$$K = [k, \infty), \text{ with } k = F_{\chi^2_{R-1}}^{-1}(1 - \alpha),$$

where $F_{\chi^2_{R-1}}^{-1}$ is the inverse of the distribution function of the χ^2–distribution with $R - 1$ degrees of freedom. For this test we can also easily compute the p–value. In our case,

$$p\text{-value} = P\left(\chi^2 > h\right) = 1 - F_{\chi^2_{R-1}}(h).$$

Example 6.4.1 We draw a sample from the normal distribution $N(10, 2)$ and then we will apply the χ^2–test to check if it is a sample from $N(10, 2)$.

MAPLE ASSISTANCE 70
```
>  X := [random[normald[10,2]](100)]:
```

We consider 12 cells.
```
>  a := min(op(X)); b := max(op(X)); R := 12:
                    a := 4.829443273
                    b := 14.27005168
>  Y := sort(X):
>  n := nops(X);
                    n := 100
>  iq := iquo(n,R);
                    iq := 8
>  lr := map(x ->x-0.01,[seq(Y[1+k*iq],k = 0..R-1),b+0.1]);
```

$lr := [4.819443273, 6.777867668, 7.474239459, 8.294541683,$
$\quad 8.806236122, 9.092604796, 9.668876136, 9.934260522,$
$\quad 10.49426606, 11.15054418, 11.46794445, 12.68285333,$
$\quad 14.36005168]$

Have a quick look at a piece of data Y.
```
>  Y[45..55];
```

$[9.342458497, 9.345649020, 9.483470065, 9.561723740, 9.678876136,$
$\quad 9.705706052, 9.790018870, 9.862590213, 9.882917617,$
$\quad 9.887176858, 9.892024672]$

```
> cells := transform[tallyinto](X, [seq(lr[k]..lr[k+1],
  k = 1..R)]);
```

$$cells := [\text{Weight}(7.474239459..8.294541683, 8),$$
$$\text{Weight}(9.668876136..9.934260522, 7),$$
$$\text{Weight}(8.294541683..8.806236122, 8),$$
$$\text{Weight}(9.934260522..10.49426606, 9),$$
$$\text{Weight}(12.68285333..14.36005168, 12),$$
$$\text{Weight}(10.49426606..11.15054418, 8),$$
$$\text{Weight}(11.15054418..11.46794445, 8),$$
$$\text{Weight}(11.46794445..12.68285333, 8),$$
$$\text{Weight}(4.819443273..6.777867668, 8),$$
$$\text{Weight}(6.777867668..7.474239459, 8),$$
$$\text{Weight}(8.806236122..9.092604796, 8),$$
$$\text{Weight}(9.092604796..9.668876136, 8)]$$

```
> w := stats[transform,frequency](cells);
```
$$w := [8, 7, 8, 9, 12, 8, 8, 8, 8, 8, 8, 8]$$
```
> lr := stats[transform,statvalue](cells);
```

$$lr := [7.474239459..8.294541683, 9.668876136..9.934260522,$$
$$8.294541683..8.806236122, 9.934260522..10.49426606,$$
$$12.68285333..14.36005168, 10.49426606..11.15054418,$$
$$11.15054418..11.46794445, 11.46794445..12.68285333,$$
$$4.819443273..6.777867668, 6.777867668..7.474239459,$$
$$8.806236122..9.092604796, 9.092604796..9.668876136]$$

```
> theor_distr := statevalf[cdf,normald[10,2]]:
> P := seq(theor_distr(op(2,lr[i])) -
  theor_distr(op(1,lr[i])), i = 1..R);
```

$$P := .0935886529, .0526383784, .0783883447, .1107081316,$$
$$.0752632094, .1198489890, .0510708140, .1415918506,$$
$$.04878773331, .04973345480, .0497301433, .1092276073$$

```
> h := add((w[i] - n*P[i])^2/(n*P[i]),i = 1..R);
```
$$h := 15.92597545$$

Fix a level of significance and find the left bound of the critical set.
```
> alpha := 0.05:
> stats[statevalf,icdf,chisquare[R - 1]](1 - alpha);
```
$$19.67513757$$

Thus we do not reject the hypothesis. Graphically, we can illustrate the critical set and the p-value.
```
> Cdf := stats[statevalf,cdf,chisquare[R-1]]:
> p_value := 1 - Cdf(h);
```
$$p\_value := .1439013414$$

```
>  pline := plots[pointplot]([[h,0],[h,Cdf(h)]],
   connect = true,linestyle = 3):
>  plots[display]([plot([Cdf, 1 - alpha],0..2*h),pline]);
```

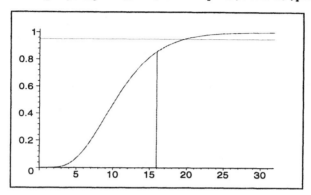

Thus, we can interpret the p–value as the smallest level of significance at which the hypothesis may be rejected using the obtained value h.

6.5 Independence Test

Assume we have two variates measured at n objects, say $x_1,\ldots,x_n$ and $y_1,\ldots,y_n$. We want to answer the question of whether or not the random variables X and Y represented by the samples are independent random variables. We will address this problem with the use of the χ^2–test for independence.

First of all, we have to make a so called contingency table. We partition the samples into R mutually disjoint cells, C_1, ..., C_R and D_1, ..., D_S respectively. Denote by $w_{i,j}$ the numbers of points of the sequence $(x_1,y_1),\ldots,(x_n,y_n)$ belonging to the cell $C_i \times D_j$. Clearly $\sum_{i,j} w_{i,j} = n$.

Let $w_{i\cdot} = \sum_j w_{i,j}$ and $w_{\cdot j} = \sum_i w_{i,j}$.

We will explore the following idea. If the random variables are independent then for each cell $C_i \times D_j$ we should have $P(C_i \times D_j) = P(C_i)P(D_j)$, which for samples would mean

$$\frac{w_{i,j}}{n} = \frac{w_{i\cdot}}{n}\frac{w_{\cdot j}}{n}.$$

Thus, for independent random variables the number

$$h = \sum_{i,j} \frac{\left(w_{i,j} - \frac{w_{i\cdot}w_{\cdot j}}{n}\right)^2}{\frac{w_{i\cdot}w_{\cdot j}}{n}}$$

should be small. It appears that under some mild assumptions the above statistic has the χ^2–distribution with the number of degrees of freedom equal to $(R-1)(S-1)$. For a given level of significance α the critical set K is the interval

$$K = [k, \infty), \text{ with } k = F^{-1}_{\chi^2_{(R-1)(S-1)}}(1-\alpha).$$

In this case,

$$p\text{-value} = P\left(\chi^2 > h\right) = 1 - F_{\chi^2_{(R-1)(S-1)}}(h).$$

Example 6.5.1 We draw two samples and then check their independence.

MAPLE ASSISTANCE 71

```
>  n := 3000:
>  X := [random[uniform[-5,5]](n)]:
>  Y := [random[normald[30,2]](n)]:
>  XY := zip((x,y) ->[x,y], X, Y):
```

We determine some intervals to serve as cells. Our choice is, to some extent, subjective. Still, we want to take into account some properties of the samples. At the same time we are establishing conditions to automatically construct the contingency table.

```
>  xm, xM := min(op(X)), max(op(X));
            xm, xM := −4.999449132, 4.998018399
>  gr1 := [seq(-5 + 2*i,i = 0..5)];
>  R := nops(gr1)-1;
            gr1 := [−5, −3, −1, 1, 3, 5]
                    R := 5
>  gr1L := [seq(gr1[i],i=1..R)]:
   gr1P := [seq(gr1[i],i=2..R+1)]:
>  a := zip((s,t)->(x->(s < x and x <= t)),gr1L,gr1P):
>  ym, yM := min(op(Y)), max(op(Y));
            ym, yM := 22.48188802, 38.08606926
>  gr2 := [21,28,30,32,39]; S := nops(gr2)-1;
            gr2 := [21, 28, 30, 32, 39]
                    S := 4
>  gr2L := [seq(gr2[i],i=1..S)]:
   gr2P := [seq(gr2[i],i=2..S+1)]:
>  b := zip((s,t)->(x->(s < x and x <= t)),gr2L,gr2P):
```

We will make the contingency table with an additional row and column for the sums $w_i.$ and $w._j$ respectively.

```
>  CT := array(1..R+1,1..S+1):
>  for i from 1 to R +1 do
   for j from 1 to S +1 do
   CT[i,j] := 0 od
   od:
```

```
>  for k from 1 to n do
   for i from 1 to R do
   for j from 1 to S do
   if (a[i](XY[k][1]) and b[j](XY[k][2])) then
   CT[i,j] := CT[i,j] +1
   fi
   od
   od
   od:
>  for i from 1 to R do
   CT[i,S+1] := add(CT[i,j],j=1..S)
   od:
   for j from 1 to S do
   CT[R+1,j] := add(CT[i,j],i=1..R)
   od:
   CT[R+1,S+1] := add(CT[R+1,j],j = 1..S):
>  print(CT);
```

$$\begin{bmatrix} 100 & 208 & 196 & 90 & 594 \\ 90 & 183 & 185 & 101 & 559 \\ 104 & 194 & 198 & 103 & 599 \\ 99 & 233 & 213 & 102 & 647 \\ 94 & 217 & 196 & 94 & 601 \\ 487 & 1035 & 988 & 490 & 3000 \end{bmatrix}$$

```
>  h := evalf(n*add(add((CT[i,j] - CT[i,S+1]*CT[R+1,j]/n)^2/
   (CT[i,S+1]*CT[R+1,j]),i = 1..R),j = 1..S));
```
$$h := 5.429582140$$

```
>  Pvalue :=
   1 - stats[statevalf,cdf,chisquare[(R-1)*(S-1)]](h);
```
$$Pvalue := .9420722267$$

This p–value strongly confirms the hypothesized independence. We also calculate the left bound of the critical at the significance level α.

```
>  alpha := 0.05:
>  stats[statevalf,icdf,chisquare[(R-1)*(S-1)]](1-alpha);
```
$$21.02606982$$

The observed value $h = 5.429582140$ is thus far away from K, which is what actually was expected.

6.6 Confidence Intervals

The following problem of estimation has important practical significance. Given a sample, estimate a particular parameter of a distribution from which the sample comes. In addition, find a set, an interval in most cases, containing the true parameter with a high probability, say probability $1 - \alpha$, where α is small. Such a set is called the *confidence interval* for the parameter and $1 - \alpha$ is called its *level of confidence*. This problem does not have a general solution, but in many particular cases there are procedures that allow us to handle it. We will consider some examples.

Suppose we want to estimate the mathematical expectation m, in particular to find its confidence interval with confidence level $1 - \alpha$, for a simple sample $x_1, \ldots, x_n$. The first task here has a simple solution: take the sample mean $\bar{x} = \frac{1}{n}\sum_{j=1}^{n} x_j$ as the estimator of m. Theoretical arguments say that this is indeed a very good estimator. We will discuss three important cases for finding the confidence interval for m, depending on what additional information we have. In all cases we are looking for the interval (a, b) having the following properties: (1) $m \in (a, b)$, (2) both a and b depend on $\bar{x}$, (3) $P_\xi(a, b) = 1 - \alpha$, where P_ξ is the unknown distribution of the random variable ξ from which the sample comes.

Case 1. The size of the sample is large ($n \geq 30$ in practice.) From the Central Limit Theorem we know that the distribution of $\bar{x}$ is $\Phi_{m, \frac{\sigma}{\sqrt{n}}}$, where σ is the standard deviation of the distribution that the sample comes from and is unknown, but from the Laws of Large Numbers may be substituted by

$$s = \sqrt{\frac{1}{n-1}\sum_{i=1}^{n}(x_i - \bar{x})^2},$$

see Problem E 4.10. We can now find a confidence interval (a, b) in a variety of ways. For example, we may want it to be symmetric about $\bar{x}$, in other words we are looking for a and b of the form: $a = \bar{x} - \varepsilon$ and $b = \bar{x} + \varepsilon$. so we have:

$$m \in (\bar{x} - \varepsilon, \bar{x} + \varepsilon) \iff \bar{x} \in (m - \varepsilon, m + \varepsilon),$$

thus

$$P_\xi(a, b) = P(\bar{x} \in (m - \varepsilon, m + \varepsilon)) = \Phi_{m, \frac{s}{\sqrt{n}}}(m + \varepsilon) - \Phi_{m, \frac{s}{\sqrt{n}}}(m - \varepsilon).$$

In particular, we want to find $\varepsilon > 0$ such that

$$\Phi_{m, \frac{s}{\sqrt{n}}}(m + \varepsilon) - \Phi_{m, \frac{s}{\sqrt{n}}}(m - \varepsilon) = 1 - \alpha.$$

Taking advantage of the properties of normal distribution we thus have:

$$2\Phi\left(\frac{\varepsilon\sqrt{n}}{s}\right) - 1 = 1 - \alpha$$

and hence

$$\varepsilon = \frac{s}{\sqrt{n}}\Phi^{-1}\left(1 - \frac{\alpha}{2}\right),$$

which is what we require.

Sometimes we would like to have a confidence interval of the form (a, ∞) or $(-\infty, b)$. Reasoning as before we can easily derive a formula for ε, namely:

$$\varepsilon = \frac{s}{\sqrt{n}} \Phi^{-1}(1 - \alpha),$$

in both cases.

Case 2. The size of the sample is small and we know that the sample comes from a normal distribution with a given standard deviation σ. In this case we know that $\bar{x}$ has the normal distribution $\Phi_{m, \frac{\sigma}{\sqrt{n}}}$, so we can derive formulas for ε as before, namely, we can have a confidence interval of the form $(\bar{x} - \varepsilon \; \bar{x} + \varepsilon)$ with

$$\varepsilon = \frac{\sigma}{\sqrt{n}} \Phi^{-1}\left(1 - \frac{\alpha}{2}\right),$$

or one sided intervals $(-\infty, \bar{x} - \varepsilon)$ or $\bar{x} + \varepsilon, \infty)$ with

$$\varepsilon = \frac{\sigma}{\sqrt{n}} \Phi^{-1}(1 - \alpha).$$

Case 3. The size of the sample is small and we know that the sample comes from a normal distribution, but we do not know standard deviation σ. In this case we use the fact that the statistic

$$t = \frac{\bar{x} - m}{s\sqrt{n}}$$

has a known distribution t_{n-1} that is called the *t–distribution* or *Student distribution with parameter* $n - 1$. The formulas for the confidence intervals have the same forms as before, but now with

$$\varepsilon = \frac{\sigma}{\sqrt{n}} t_{n-1}^{-1}\left(1 - \frac{\alpha}{2}\right),$$

for a symmetric interval and

$$\varepsilon = \frac{\sigma}{\sqrt{n}} t_{n-1}^{-1}(1 - \alpha)$$

for the one sided intervals.

If the sample is small and we do not have any information about the underlying distribution, then the situation is much more complicated. In many cases, however, the assumption of normality is not important and the formulas established above for the confidence intervals may still be used.

We may want to estimate some other parameter on the basis of a sample. The procedure is basically the same. First, we have to decide which estimator (statistic) should be used. Then, we should find a statistic with a known distribution that is appropriate for determining a confidence interval for a given confidence level. Remember, that in most cases there are various types of confidence intervals, so we need to decide which one we really need.

Note that estimation and hypothesis testing are closely related to each other in the sense that they use the same mathematical tools and similar philosophy, but handle quite different types of problems.

6.7 Inference for Numerical Simulations

In the previous section we have assumed we were given a simple sample and then on its basis we have performed a test or an estimation. Yet, in numerical simulations we can use samples of sizes, practically, as large as we wish. In such cases we can apply a bit different approach than we used before. We explain some common idea in the case of determining confidence intervals for the mean.

Example 6.7.1 Let ξ be a random variable, which can be simulated on a computer and assume that we want to find a confidence interval for the mean $\mathbb{E}(\xi)$. This problem is quite important when we are not able to find a formula for the distribution function of ξ. As an example take $\xi = \eta + 3\sin(\eta)$, where η is random variable exponentially distributed with parameter λ, say $\lambda = 0.5$.

To solve this problem we fix a number K and define the average:

$$\bar{\xi} = \frac{1}{K}(\xi_1 + \ldots, \xi_K),$$

where $\xi_1, \ldots, \ldots, \xi_K$ are i.i.d. random variables with common distribution P_ξ. Certainly, $\mathbb{E}(\bar{\xi}) = \mathbb{E}(\xi)$. The advantage of considering $\mathbb{E}(\bar{\xi})$ is, that for large K, practically $K \geq 100$ (in many case $K \geq 20$ is good enough) $\bar{\xi}$ is almost normally distributed by the Central Limit Theorem. Now, we can draw a simple sample from $\bar{\xi}$ and on its basis determine confidence intervals of its mean using formulas derived in previous section. Turn back to the specific random variable mentioned above.

MAPLE ASSISTANCE 72

Define a procedure which simulates random variable ξ.
```
> with(stats):
> randomdata := proc(n)
  map(x->x+3*sin(x), [random[exponential[0.5]](n)])
  end:
```

Compare samples from ξ and from $\bar{\xi}$.
```
> L := 100: K := 100:
> listX := NULL: listY := NULL:
> from 1 to L do
```

```
>  X := randomdata(K);
>  Y := evalf(describe[mean](X));
>  listX := listX,op(X):
   listY := listY,Y:
   od:
>  statplots[histogram]([listX],
   numbars = ceil(log[2](L*K)),area=count);
```

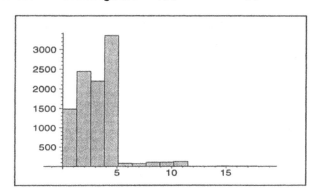

```
>  statplots[histogram]([listY],
   numbars = ceil(log[2](L)),area = count);
```

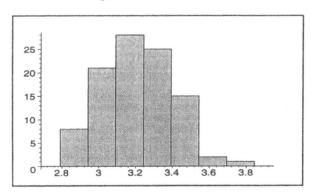

We will check graphically normality of the sample from $\bar{\xi}$ plotting the density function of corresponding normal distribution along the histogram, compare with page 155. (We urge the reader to perform a more advanced test χ^2.) Incidentally, we get the empirical mean.

```
>  m := evalf(describe[mean]([listY]));
   sigma := evalf(describe[standarddeviation]([listY]));
   c := 3.486*sigma: h := evalf(c/L^(1/3)):
```

$$m := 3.210765118$$

$$\sigma := .1953830712$$

```
>  k := floor(3*sigma/h):
   a := m - (k+0.5)*h: b := m + (k+0.5)*h:
   1Y := transform[tallyinto['outliers']]([listY],
   [seq(a+i*h..a+(i+1)*h,i=0..2*k)]):
>  1Y2 := transform[scaleweight[1/L]](1Y):
>  outliers;
```

$$[3.844601277]$$

```
>  hist := statplots[histogram](1Y2,colour=cyan):
>  pp:=plot(stats[statevalf,pdf,normald[m,sigma]],a..b ,
   color = black):
>  plots[display](hist,pp);
```

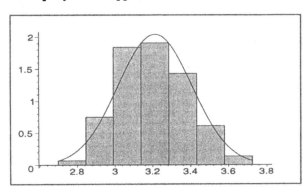

We make a procedure which will compute confidence intervals.

```
>  conf_intervals_mean := proc(L,K,alpha)
   local listX, listY, X,Y, mY, sigmaY, t, epsilon:
   listX := NULL: listY := NULL:
>  from 1 to L do
>  X := randomdata(K);
>  Y := evalf(describe[mean](X));
>  listX := listX,op(X):
   listY := listY,Y:
   od:
   mY := evalf(describe[mean]([listY]));
   sigmaY := evalf(describe[standarddeviation]([listY]));
   if L < 30 then
   t := statevalf[icdf,studentst[L-1]](1 - alpha/2/100)
   else
   t := statevalf[icdf,normald](1 - alpha/2/100)
   fi;
   epsilon := evalf(sigmaY*t/sqrt(L)):
   [mY - epsilon, mY + epsilon]
   end:
```

Compute some confidence intervals.

```
>  to 5 do
   conf_intervals_mean(20,100,5)
   od;
```

$$[3.154995248, 3.308047640]$$

$$[3.110505114, 3.250267130]$$

$$[3.061932207, 3.236378181]$$

$$[3.150887309, 3.298534027]$$

$$[3.168553474, 3.349749746]$$

```
>  to 5 do
   conf_intervals_mean(100,100,5)
   od;
```

$$[3.172364565, 3.254044501]$$

$$[3.232558819, 3.302051101]$$

$$[3.131370825, 3.203610263]$$

$$[3.161488080, 3.240110820]$$

$$[3.137129082, 3.215572044]$$

Note a paradox: two of the above fives confidence intervals are disjoint.

Finally, it is instructive to compute the actual mean.

```
>  int((x+3*sin(x))*0.5*exp(-0.5*x),x = 0..infinity);
```
$$3.200000000$$

6.8 MAPLE Session

M 6.1 *Find the prime factors of the numbers a and p used by* MAPLE *in procedure* **rand**. *Are these numbers really relatively prime?*

```
>  a := 427419669081: p := 999999999989:
>  ifactor(a); ifactor(p);
```
$$(3) (61) (491) (4756877)$$
$$(999999999989)$$

Hence p is a prime number and thus a and p are, clearly, relatively prime.

M 6.2 *We present a graphical method to check if a sample is simple.*

We will plot pairs of successive numbers drawn according to the $U(0,1)$ distribution. If the resulting points fill the unit square evenly we have no evidence against the randomness of the numbers.

```
>  U := [random[uniform](2000)]:
>  W := [seq([U[2*i-1],U[2*i]],i = 1..1000)]:
>  plots[pointplot](W, symbol = CIRCLE);
```

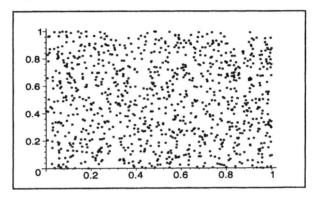

M 6.3 *We generate numbers belonging to the interval $[0,1]$ using the following procedure and then check if they can be treated as random numbers.*

We fix some x_0 such that $0 < x_0 < 1$ and define successive points using the formula

$$x_{n+1} = 4x_n(1 - x_n), \quad n = 0, 1, 2, \ldots.$$

Next, we transform the numbers using the map

$$F(x) = \frac{\arcsin(2x - 1)}{\pi} + \frac{1}{2}.$$

We define an appropriate procedure and then generate 1500 numbers.

```
>  rrand := proc(x0,n)
   local alist, x, F:
   alist := x0: x := x0:
   to 20 + n do
   x := 4*x*(1-x):
   alist := alist,x
   od:
```

```
    F := x -> evalf(arcsin(2*x-1)/Pi +1/2):
    map(F,[alist[21..20+n]])
    end:
>   X := rrand(0.23243,1500):
```

Examine a piece of data.
```
>   X[600..610];
```

[.3900839948, .7801679895, .4396640208, .8793280419, .2413439165,
 .4826878328, .9653756645, .0692486688, .1384973372,
 .2769946743, .5539893484]

Construct the histogram, compute the mean and the variance, remembering
that for the $U(0,1)$ distribution they should be 0.5 and 0.085 respectively.
```
>   with(stats):
>   statplots[histogram](X, area = count);
```

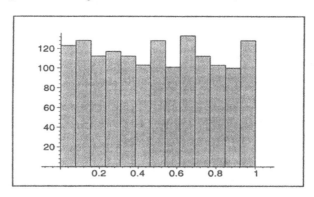

```
>   with(describe):
>   mX := mean(X);
>   D2X := variance(X);
```
$$mX := .4938188054$$
$$D2X := .08524044160$$

Check if the sample is simple using the graphical method.
```
>   W := [seq([X[2*i-1],X[2*i]],i = 1..750)]:
>   plots[pointplot](W, symbol = CIRCLE);
```

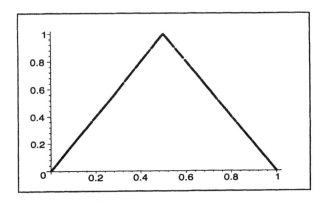

The above should be a strong warning to the reader.

M 6.4 *We will explore graphically the t–distribution and compare it with the standard normal distribution. We use different values of the parameter n, called degrees of freedom.*

```
>  with(stats):
>  with(plots):
>  display([seq( plot([statevalf[pdf,studentst[n]](x),
   statevalf[pdf,normald]](x), x = -3..3, color = [BLUE,RED],
   thickness = [1,2], tickmarks=[2,0]),
   n = [1,2,3,5,10,15,20,25,30] )],insequence = true):
>  display(%);
```

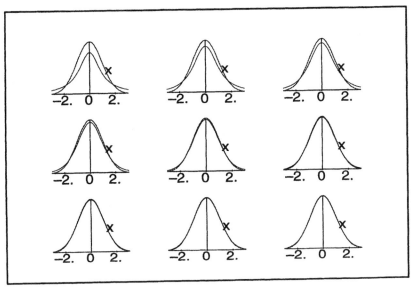

M 6.5 *We will check experimentally formulas for the size of symmetric confidence intervals when the sample is small.*

We write a procedure which K times draws a sample of size N from the $\Phi_{m,\sigma}$ distribution and each time computes the size of the symmetric confidence interval $(\bar{x} - \varepsilon, \bar{x} + \varepsilon)$ using both normal and t–distribution and computes also the distance between theoretical and experimental means. As a result we are given a K element list of the triples. The last parameter of the procedure is α, the significance level.

```
>  conf_interval := proc(K,N,m,sigma, alpha)
>  local alist, X, mX,s, epsilon1, epsilon2;
>  alist := NULL:
>  to K do
>  X := [random[normald[m,sigma]](N)]:
>  mX := describe[mean](X):
>  s := describe[standarddeviation[1]](X):
>  epsilon1 := evalf(sigma/sqrt(N)*statevalf[icdf,
     normald](1 - alpha/2)):
>  epsilon2 := evalf(s/sqrt(N)*statevalf[icdf,
     studentst[N-1]](1 - alpha/2)):
>  alist := alist,[epsilon1,epsilon2,abs(mX - m)];
     od:
>  alist:
     end:
```

We run the procedure.

```
>  conf_interval(10,20,30,4,0.1);
```

$$[1.471201810, 1.659498222, .49610198],$$
$$[1.471201810, 1.544641144, 1.82998811],$$
$$[1.471201810, 1.540593583, .53867647],$$
$$[1.471201810, 1.710287930, .58704233],$$
$$[1.471201810, 1.852954221, .19811139],$$
$$[1.471201810, 1.346692438, .70179917],$$
$$[1.471201810, 1.736842683, .39601310],$$
$$[1.471201810, 1.258580144, .96855298],$$
$$[1.471201810, 1.738690159, .15854762],$$
$$[1.471201810, 1.817864263, .09693126]$$

We see by inspection that in 9 of 10 experiments the confidence intervals covered the true value of m. However, in the second experiment both of the confidence intervals were too short to cover the true value of the mean.

We perform 100 experiments with samples of size 10 with confidence level 0.95 this time. We list all of the cases when confidence interval determined by means of normal or Student distribution does not cover the actual mean.

```
>  f := x -> (x[3] < min(x[1],x[2])):
>  remove(f,[conf_interval(100,10,30,4,0.05)]);
```

$$[[2.479180130, 2.689284995, 2.74031416],$$
$$[2.479180130, 2.666510808, 3.12648738],$$
$$[2.479180130, 3.218988376, 3.22663800],$$
$$[2.479180130, 1.646172461, 1.84898106],$$
$$[2.479180130, 2.055463315, 2.80928558],$$
$$[2.479180130, 2.080460763, 3.31815042],$$
$$[2.479180130, 3.597143822, 2.75984854]]$$

We will perform 1000 similar experiments with other sample sizes and confidence levels.

```
>  nops(remove(f,[conf_interval(1000,5,30,4,0.05)]));
                        84

>  nops(remove(f,[conf_interval(1000,15,30,4,0.1)]));
                        133
```

6.9 Exercises

E 6.1 Use $p = 101$, $a = 999$ and $b = 0$ to define a linear congruential pseudo-random number generator. Draw 1000 numbers using this generator and then test their statistical properties.

E 6.2 Modify the above procedure rrand, see M 6.3, by skipping every two numbers from the sequence x_n. Is this a good generator?

E 6.3 Search MAPLE Help for procedures that can be used to generate a sample from the normal distribution.

E 6.4 Generate a sample of size 100 from the Gompertz distribution. The density of this distribution is zero for negative x and for positive x is defined by

$$f(x) = h(x) \exp\left(-\int_0^x h(y)\,dy\right),$$

where $h(x) = \exp(\alpha x + \beta)$ is the so called hazard function. Put $\alpha = 0.06$ and $\beta = -7$. The Gompertz distribution is used to describe life expectancy.

E 6.5 Draw a 100 element sample from the normal distribution $N(20, 2)$.

(a) Check the hypothesis that the mean is 21;

(b) Check the hypothesis that the sample comes from the distribution $N(20, 3)$.

E 6.6 Check the hypothesis that the sample X in M 6.3 has $U(0, 1)$ distribution.

E 6.7 We are given the numbers

$$5.37, 6.51, 5.99, 5.63, 5.73, 5.96, 6.04.$$

Knowing that the numbers are randomly chosen and come from a normal distribution check the hypothesis that the mean of the distribution is $m = 6$. Find confidence intervals for m. Consider three different types of a critical set and a confidence interval.
   Hint. Use the t–distribution.

E 6.8 Establish an equivalent definition of p–value in terms of confidence intervals.

E 6.9 Perform a computer experiment to check the necessity of the assumption of normality of a sample if one determines confidence intervals for the mean.
   Hint. Modify procedure conf_interval to draw a sample from various distributions.

E 6.10 Let X and Y be independent random variables with the same $N(10, 3)$ distribution. Define $Z = 0.1X + 2Y$. Generate a sample of size 100 to check whether X and Z are independent using (a) the independence test, and (b) the correlation coefficient.

7 Stochastic Processes

A sequence of random variables $X_1, X_2, \ldots, X_n, \ldots$ often describes, or may be conveniently imagined to describe, the evolution of a probabilistic system over discrete instants of time $t_1 < t_2 < \cdots < t_n < \cdots$. We then say that it is a *discrete–time stochastic process*. Stochastic processes may also be defined for all time instants in a bounded interval such as $[0, 1]$ or in an unbounded interval such as $[0, \infty)$, in which case we call them *continuous–time stochastic processes*.

In this chapter we will present some basic concepts from the theory of stochastic processes, in particular those that are needed for an exploration of stochastic differential equations. In the first section, however, we return to the concept of conditional probability and use a generalization of it to define conditional expectation; both of these concepts are useful in expressing the probabilistic relationship between the random variables at different times that form a stochastic process. Then, in the second section, we develop a simple but nevertheless important class of discrete–time processes known as Markov chains, before considering some fundamental continuous–time processes such as Wiener and Poisson processes in the remaining sections of the chapter.

7.1 Conditional Expectation

In Chapter 1 we defined the conditional probability $P(B|A)$ assuming that the probability $P(A)$ of the conditioning event A is positive. We also observed in Example 1.4.5 that such a probability also makes sense in some circumstances when $P(A) = 0$. In this section we define conditional probability of a given set with respect to a given σ-algebra, which extends the previous standard definition and, in particular, covers the case described in Example 1.4.5. We will go further and define the conditional expectation of a given random variable with respect to a given σ-algebra.

Let (Ω, Σ, P) be a probability space, let $\mathcal{A} \subset \Sigma$ be a σ-algebra and let $B \in \Sigma$.

Definition 7.1.1 *A function $\varphi : \Omega \longrightarrow \mathbb{R}$ is said to be a representative of conditional probability of the event B with respect to $\mathcal{A}$ if:*

(1) φ is measurable with respect to $\mathcal{A}$, or briefly $\mathcal{A}$-measurable; and

(2) for any $A \in \mathcal{A}$

$$\int_A \varphi \, dP = P(B \cap A). \tag{7.1}$$

It can be shown that there is at least one such representative function φ and that if ψ also satisfies conditions (1) and (2), then $\varphi = \psi$, w.p.1. The family of all the functions satisfying (1) and (2) is denoted by $P(B|\mathcal{A})$ and is called the *conditional probability of B with respect to $\mathcal{A}$*. We usually identify one of the representatives of the conditional probability with the conditional probability itself.

Example 7.1.1 Let $A_1, \ldots, A_k \in \Sigma$ be pairwise disjoint with $\bigcup_{i=1}^{k} A_i = \Omega$ and let $\mathcal{A} = \sigma(A_1, \ldots, A_k)$ denote the smallest σ-algebra containing all of the A_i. Finally, let $B \in \Sigma$ and define:

$$\varphi(\omega) = \begin{cases} \dfrac{P(B \cap A_i)}{P(A_i)} & \text{if } \omega \in A_i \text{ and } P(A_i) > 0, \\ c & \text{if } \omega \in A_i \text{ and } P(A_i) = 0, \end{cases}$$

where $c \in \mathbb{R}$ is an arbitrary constant. The function φ is $\mathcal{A}$-measurable because it is constant on each A_i. In addition, condition (7.1) holds since any $A \in \mathcal{A}$ is a finite union of some A_i's, say $A = A_{i_1} \cup \ldots \cup A_{i_s}$, and both sides in (7.1) are equal to:

$$P((B \cap A_{i_1}) \cup \ldots \cup (B \cap A_{i_s})).$$

Example 7.1.2 Let $\Omega = \mathbb{R}^2$ and let Σ be the σ-algebra $\mathcal{B}(\mathbb{R}^2)$ of all Borel subsets of the plane $\mathbb{R}^2$. Then $\mathcal{A} = \{A \times \mathbb{R} : A \in \mathcal{B}(\mathbb{R})\}$, where $\mathcal{B}(\mathbb{R})$ is the σ-algebra of all Borel subsets of the line $\mathbb{R}$, is clearly is a σ-algebra contained in Σ. Finally, consider the event $\mathbb{R} \times B$ for a fixed set $B \in \mathcal{B}(\mathbb{R})$. We can interpret this as a two step experiment: Choosing $\mathcal{A}$ provides us with information about the first step of the experiment, then choosing the event $\mathbb{R} \times B$ means we are interested in the outcome of the second step.

Assume first that P is a discrete distribution on $\mathbb{R}^2$, i.e. there are distinct points $(x_i, y_j) \in \mathbb{R}^2$ and the probabilities $P(x_i, y_j) = p_{ij}$. Then for any $(x, y) \in \mathbb{R}^2$ define

$$\varphi(x, y) = \frac{\sum\limits_{j: y_j \in B} p_{ij}}{\sum\limits_{j} p_{ij}} \tag{7.2}$$

if x is equal to some x_i and $\sum_j p_{ij} > 0$, and define $\varphi(x, y) = c$, where c is an arbitrary constant, for other pairs (x, y).

Now assume that $f : \mathbb{R}^2 \longrightarrow \mathbb{R}$ is the density of the distribution P and for any $(x, y) \in \mathbb{R}^2$ define

$$\varphi(x,y) = \begin{cases} \dfrac{\int_B f(x,y)\,dy}{\int_{\mathbb{R}} f(x,y)\,dy} & \text{if } \int_{\mathbb{R}} f(x,y)\,dy > 0, \\ c & \text{if } \int_{\mathbb{R}} f(x,y)\,dy = 0, \end{cases} \tag{7.3}$$

where again $c \in \mathbb{R}$ is an arbitrary constant.

In fact, in both cases, φ is $\mathcal{A}$-measurable since its values do not depend on the second variable y and it is expressed in terms of measurable functions of x. In addition, condition (7.1) holds: by the Total Probability Law in the discrete case and by Fubini's Theorem, Theorem 2.3.4, in the continuous case. Namely in the latter case we have:

$$\int_{A \times \mathbb{R}} \varphi\,dP = \int_{A \times \mathbb{R}} \varphi(x,y) f(x,y)\,d(x,y) = \int_{A \times B} f(x,y)\,d(x,y),$$

which is just $P((A \times \mathbb{R}) \cap (\mathbb{R} \times B))$, as required. Thus, φ defined by (7.2) and (7.3) are representatives of corresponding conditional probabilities.

For the particular σ-algebra above we will often write $P(B|x)$ instead of $P(B|\mathcal{A})$.

Note that formula (7.2) could in fact also have been derived from the previous example. Besides, formula (7.3) extends (7.2).

Example 7.1.3 We can now correctly reformulate the situation described in Example 1.4.5. Recall that we chose a number x at random from the interval $[0,1]$ and then a number y at random from the interval $[0,x]$. We want to determine $P\left(y \leq \frac{1}{3} \middle| x = \frac{1}{2}\right)$, but first we have to find the probability distribution in our experiment. The uniform distribution on $[0,1]$ provides the natural density here, specifically

$$f(x,y) = \begin{cases} \dfrac{1}{x} & \text{if } 0 < y < x < 1, \\ 0 & \text{otherwise.} \end{cases}$$

Then, using formula (7.3), we can easily determine the conditional probability $P(B|x)$ or, to be more precise, one of its representatives

$$\varphi(x,y) = \begin{cases} 1 & \text{if } 0 < x < \frac{1}{3}, \\ \dfrac{1}{3x} & \text{if } \frac{1}{3} \leq x \leq 1 \\ 0 & \text{otherwise.} \end{cases}$$

Hence, in particular, $P(B|\frac{1}{2}) = \varphi(\frac{1}{2}, y) = \frac{2}{3}$, as expected.

Example 7.1.4 In the next sections we will often encounter expressions of the form

$$P(Y \in B|X_1 = x_1, \ldots X_k = x_k),$$

where B is a Borel set, $Y, X_1, \ldots, X_k$ are random variables and $x_1, \ldots, x_k$ are points. When the random vector $(Y, X_1, \ldots, X_k)$ has a discrete or a continuous distribution, we will interpret such expressions in the manner of Example 7.1.2. Specifically, in the discrete case:

$$P(X \in B | Y_1 = y_1, \ldots Y_k = y_k) = \frac{\displaystyle\sum_{j : y_j \in B} P(X_1 = x_1, \ldots, X_k = x_k, Y = y_j)}{\displaystyle\sum_j P(X_1 = x_1, \ldots, X_k = x_k, Y = y_j)},$$

and in the continuous case:

$$P(X \in B | Y_1 = y_1, \ldots Y_k = y_k) = \frac{\displaystyle\int_B f(x_1, \ldots, x_k, y)\, dy}{\displaystyle\int_{\mathbb{R}^k} f(x_1, \ldots, x_k, y)\, dy},$$

when the denominators are positive, and arbitrary constants otherwise.

We will now introduce the concept of conditional expectation for such conditional probabilities. First, we recall that the mathematical expectation of a random variable extends the concept of probability of an event in the sense that if A is an event with indicator function I_A, then $\mathbb{E}(I_A) = P(A)$. The following definition is based on a direct generalisation of this to a conditional probability with respect to σ-algebra.

Let (Ω, Σ, P) be a probability space, let $\mathcal{A} \subset \Sigma$ be a σ-algebra and let $\xi : \Omega \longrightarrow \mathbb{R}$ be a random variable.

Definition 7.1.2 *A function* $\varphi : \Omega \longrightarrow \mathbb{R}$ *is said to be a representative of the conditional expectation of the random variable* ξ *with respect to* $\mathcal{A}$, *if:*
(1) φ *is* $\mathcal{A}$*-measurable,*
(2) For any $A \in \mathcal{A}$

$$\int_A \varphi\, dP = \int_A \xi\, dP. \tag{7.4}$$

When $\mathbb{E}(\xi) < \infty$, it can be shown that there is at least one such function φ and that if ψ also satisfies conditions (1) and (2), then $\varphi = \psi$, w.p.1. The family of all functions satisfying (1) and (2) is denoted by $\mathbb{E}(\xi | \mathcal{A})$ and is called the *conditional expectation of* ξ *with respect to* $\mathcal{A}$. Here too, we usually identify the representatives of the conditional expectation with the conditional expectation itself.

Example 7.1.5 Let $A_1, \ldots, A_k \in \Sigma$, be pairwise disjoint with $\bigcup_{i=1}^k A_i = \Omega$ and let $\mathcal{A} = \sigma(A_1, \ldots, A_k)$ be the smallest σ-algebra containing the all A_i. Finally, let ξ be a random variable on Ω and define:

$$\varphi(\omega) = \begin{cases} \dfrac{\int_{A_i} \xi\, dP}{P(A_i)} & \text{if } \omega \in A_i \text{ and } P(A_i) > 0, \\ c & \text{if } \omega \in A_i \text{ and } P(A_i) = 0, \end{cases}$$

where c is an arbitrary constant. This is obviously (a representative of) the required conditional expectation $\mathbb{E}(\xi|\mathcal{A})$.

Quite often a σ-algebra $\mathcal{A}$ is generated by a random variable η, that is, $\mathcal{A}$ is the smallest σ-algebra such that η is $\mathcal{A}$-measurable, which will be denoted by $\mathcal{A} = \sigma(\eta)$. Define $\mathbb{E}(\xi|\eta) = \mathbb{E}(\xi|\sigma(\eta))$ and write $\mathbb{E}(\xi|\eta = x) = \mathbb{E}(\xi|\eta)(\omega)$ when $x = \eta(\omega)$. The latter can be interpreted as the (usual) expectation of ξ if one knows that $\eta = x$.

Example 7.1.6 Let (η, ξ) be a 2 two dimensional random vector with a continuous probability distribution with density f. For any $\omega \in \Omega$ and (x, y) $\in \mathbb{R}^2$ with $x = \eta(\omega)$ define:

$$\varphi(\omega) = \begin{cases} \dfrac{\int_{\mathbb{R}} y f(\eta(\omega), y)\, dy}{\int_{\mathbb{R}} f(\eta(\omega), y)\, dy} & \text{if } \int_{\mathbb{R}} f(\eta(\omega), y)\, dy > 0, \\ c & \text{if } \int_{\mathbb{R}} f(\eta(\omega), y)\, dy = 0, \end{cases} \qquad (7.5)$$

where $c \in \mathbb{R}$ is an arbitrary constant. This function φ is $\mathcal{A}$-measurable and satisfies condition (7.4) by Fubini's Theorem, Theorem 2.3.4, so we can write

$$\mathbb{E}(\xi|\eta = x) = \begin{cases} \dfrac{\displaystyle\int_{\mathbb{R}} y f(x, y)\, dy}{\displaystyle\int_{\mathbb{R}} f(x, y)\, dy} & \text{if } \int_{\mathbb{R}} f(x, y)\, dy > 0, \\ c & \text{if } \int_{\mathbb{R}} f(x, y)\, dy = 0. \end{cases} \qquad (7.6)$$

The above formula can be immediately extended for any random vector to a case, when η is a random vector.

We will now state some basic properties of the conditional expectation, the proofs of which are straightforward consequences of the definitions and the corresponding properties for integrals.

Theorem 7.1.1 *Let* (Ω, Σ, P) *be a probability space, let* ξ *be a random variable defined on* Ω *with finite expectation* $\mathbb{E}(\xi)$, *and let* $\mathcal{A} \subset \Sigma$ *be a* σ-algebra. *Then*

1. *If* $\mathcal{A} = \Sigma$, *then* $\mathbb{E}(\xi|\mathcal{A}) = \xi$.
2. *If* $\mathcal{A} = \{\emptyset, \Omega\}$, *then* $\mathbb{E}(\xi|\mathcal{A}) = \mathbb{E}(\xi)$.
3. $\mathbb{E}(\mathbb{E}(\xi|\mathcal{A})) = \mathbb{E}(\xi)$.
4. *If* ξ *is* $\mathcal{A}$-measurable, *then* $\mathbb{E}(\xi|\mathcal{A}) = \xi$.
5. *If* ξ_1, ξ_2 *have finite expectations and* $a \in \mathbb{R}$, *then*

$$\mathbb{E}(\xi_1 + \xi_2|\mathcal{A}) = \mathbb{E}(\xi_1|\mathcal{A}) + \mathbb{E}(\xi_2|\mathcal{A}) \quad \text{and} \quad \mathbb{E}(a\xi|\mathcal{A}) = a\mathbb{E}(\xi|\mathcal{A}).$$

6. If $\xi_1 \leq \xi_2$, then $\mathbb{E}(\xi_1|\mathcal{A}) \leq \mathbb{E}(\xi_2|\mathcal{A})$.
7. If ζ is $\mathcal{A}$-measurable, then $\mathbb{E}(\zeta\xi|\mathcal{A}) = \zeta\mathbb{E}(\xi|\mathcal{A})$.

7.2 Markov Chains

Let $\mathbf{P} : E \times E \longrightarrow \mathbb{R}$ and $\mathbf{p} : E \longrightarrow \mathbb{R}$, where $E \subset \mathbb{R}^d$ be a finite or countable set. We can consider $\mathbf{P}$ and $\mathbf{p}$ as being a finite or infinite matrix and a column vector, respectively, with nonegative components $\mathbf{P}(i,j)$ and $\mathbf{p}(i)$, $i, j \in E$.

Definition 7.2.1 *A sequence $\{X_n\}_{n\geq 0}$ of random vectors that are defined on a common probability space (Ω, Σ, P) and take values in $\mathbb{R}^d$ is called a Markov chain if the following conditions hold:*

1. $P(X_0 = i) = \mathbf{p}(i)$ *for each* $i \in E$;
2. *for each* $n \geq 0$ *and every sequence* $i_0, \ldots, i_{n+1} \subset E$

$$P(X_{n+1} = i_{n+1}|(X_0 = i_0, \ldots, X_n = i_n))$$

$$= P(X_{n+1} = i_{n+1}|X_n = i_n) = \mathbf{P}(i_n, i_{n+1});$$

3. $\sum_{i\in E} \mathbf{p}(i) = 1$;

4. $\sum_{j\in E} \mathbf{P}(i,j) = 1$ *for every* $i \in E$.

We can interpret the above conditions as follows. Consider E as a collection of all possible states of some system. Then X_n indicates which state is active at the n-th instant of time. Thus, we assume that the system at time zero is in state i with probability $\mathbf{p}(i)$ and then moves according to the rule that the probability of transition from some state to another depends only on these states and not on the previous states that have been visited or on the particular instant of time n. Property 3 says that the system starts in E at time zero with probability one. In fact,

$$P(X_0 \in E) = \sum_{i\in E} \mathbf{p}_i = 1.$$

By Property 4 we also have

$$P(X_{n+1} \in E|X_n = i) = \sum_{j\in E} P(X_{n+1} = j|X_n = i) = \sum_{j\in E} \mathbf{P}(i,j) = 1$$

for any $i \in E$, which means that it is not possible to leave the set E at any time. We call $\mathbf{p}$ an initial distribution and $\mathbf{P}$ a transition matrix.

We will now describe some classical examples of Markov chains, beginning with the simplest case of random walks.

7.2.1 Random Walks

We start with the simplest random walk.

Example 7.2.1 Imagine that a particle moves along the real line in the following manner. At time zero the particle is at the origin and at the following instances of time it moves one unit to the right or to the left with probabilities p and $q = 1 - p$, respectively. If $p = q = \dfrac{1}{2}$ we have a *standard random walk*.

To see that this situation is a Markov chain let the set of all states be the set of integers $E = \mathbb{Z} \subset \mathbb{R}$ and let X_n be the position of the particle at time n. Also define

$$\mathbf{p}(i) = \begin{cases} 1 & \text{for } i = 0, \\ 0 & \text{for } i \neq 0, \end{cases}$$

and

$$\mathbf{P}(i,j) = \begin{cases} q & \text{for} & j = i - 1, \\ p & \text{for} & j = i + 1, \\ 0 & \text{otherwise.} \end{cases}$$

The above random walk can be modified in various ways. For example, assume in addition that the particle can remain in each state with a probability r, so now $p + q + r = 1$. Another modification is to introduce a barrier or two which the particle is not allowed to pass across. In such a situation the set of states E may be finite when two barriers are present. For example, if we fix the barriers at points, A and B with $A < 0 < B$, then the set E contains $B - A + 1$ states and the $(B - A + 1) \times (B - A + 1)$ matrix $\mathbf{P}$ be as follows:

$$\mathbf{P} = \begin{bmatrix} s_a & 1 - s_a & 0 & \cdots & & \cdots & 0 \\ 0 & q & r & p & 0 & & \cdots \\ \cdots & \cdots & \cdots & \cdots & \cdots & \cdots & \cdots \\ \cdots & \cdots & \cdots & \cdots & \cdots & \cdots & \cdots \\ \cdots & 0 & q & r & p & 0 \\ 0 & \cdots & & \cdots & 0 & 1 - s_b & s_b \end{bmatrix},$$

where s_a and s_b denote the probabilities that the particle is absorbed by the barriers at A and B, respectively. Two interesting extreme cases are when these probabilities are 0 (perfect elasticity of the barriers) or 1 (full absorption).

We demonstrate some examples using MAPLE.

MAPLE ASSISTANCE 73

We define a procedure which will generate plots of a path of a given length beginning at zero for a random walk with specified parameters.

```
> with(stats):
> RandWalk := proc(N,p,q,sa,sb,M)
> local A, B, r, PA, PC, PB, state, step, path:
> A := -N: B := N: r := 1 - p - q:
> PA := empirical[0.,sa,1-sa]:
  PC := empirical[q,r,p]:
  PB := empirical[1-sb,sb,0.]:
> state := 0: path := state:
> from 1 to M do
  if (A < state and state < B) then
  step := trunc(stats[random,PC](1))-2
  elif state = A then step := trunc(stats[random,PA](1))-2
  elif state = B then step := trunc(stats[random,PB](1))-2
  fi:
  state := state + step:
  path := path,state
  od:
> plots[pointplot]([seq([i,path[i]],i = 1..nops([path]))],
  axes=BOXED):
  end:
```

First, consider a short path of 20 points with close barriers.

```
> RandWalk(3,0.5,0.5,0.8,0.8,20);
```

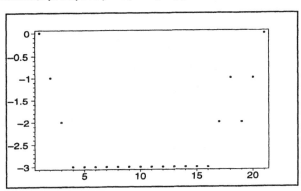

Now, we generate four paths of another random walk.

```
> plots[display]([seq(
> RandWalk(10,0.45,0.5,0.8,0.8,300),k = 1..4)],
  insequence= true):
> plots[display](%);
```

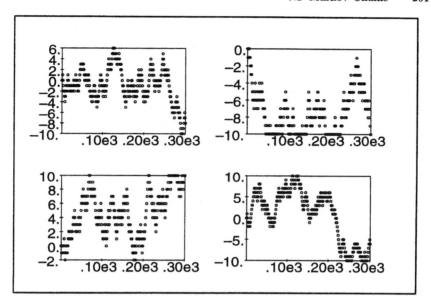

$$\cdots$$

We can describe a random walk (with no barriers for simplicity) in a slightly different but useful setting.

Example 7.2.2 Suppose that the particle starts at some integer point i, so $X_0 = i$, and that

$$X_{n+1} = X_n + \xi_{n+1} \quad \text{for } n = 0, 1, 2, \ldots,$$

where ξ_1, ξ_2, ξ_3, ... are independent random variables taking values -1, 0 and 1 with probabilities p, q and r, respectively.

We can also consider random walks in the plane and, even more generally, in d-dimensional space.

Example 7.2.3 For simplicity let $p = q = \frac{1}{2}$, so $r = 0$. Suppose that $X_0 = i \in \mathbb{Z}^d$ and that

$$X_{n+1} = X_n + \xi_{n+1} \quad \text{for } n = 0, 1, 2, \ldots,$$

where ξ_1, ξ_2, ξ_3, ... are independent random vectors taking 2^d values $(\varepsilon_1, \ldots, \varepsilon_d) \in \{-1, 1\}^d$, so $\varepsilon_j = \pm 1$, each with probability $\dfrac{1}{2^d}$.

Let us note that each component of d-dimensional random walk is in fact a one dimensional standard random walk and these random walks are independent.

We now present some examples of what seem to be other types of Markov chains but are in fact equivalent to random walks.

Example 7.2.4 Suppose that Alice and Barbara have financial capital A and B (in dollars), respectively. They repeat a game (chess, for example) with the loser paying $1 to the winner after each game until one of them become bankrupt. Let the probabilities of success be p for Alice and q for Barbara with $r = 1 - p - q \geq 0$ being the probability of tie, and let X_n denote Alice's capital after the n-th game. The sequence of games here is in fact equivalent to a one dimensional random walk starting at the point A and having two absorbing barriers at points 0 and $A + B$.

Example 7.2.5 Two boxes contain k balls each, where k of them are white and k of them are red. At successive instances of time two balls are drawn, one ball from each box, and then put into the other box. Let X_n denote the number of white balls in the first box (thus also the number of red balls in the other box) at instant n. The X_n's here thus form a Markov chain, which is also equivalent to a one dimensional random walk with the components of the transition matrix $\mathbf{P}$ being all zero except for

$$\mathbf{P}(i, i - 1) = \left(\frac{i}{k}\right)^2, \quad \mathbf{P}(i, i + 1) = \left(\frac{k - i}{k}\right)^2, \quad \mathbf{P}(i, i) = \frac{2(k - i)i}{k^2}.$$

7.2.2 Evolution of Probabilities

The first step in investigating a Markov chain is finding a formula for the distributions of each of the random variable X_n, that is a column probability vector $\mathbf{p}_n$ with

$$\mathbf{p}_n(j) = P(X_n = j)$$

for each $n \geq 1$ and $j \in E$. Applying the Law of Total Probability we obtain

$$\mathbf{p}_n(j) = P(X_n = j)$$
$$= \sum_{i \in E} P(X_n = j | X_{n-1} = i) P(X_{n-1} = i) = \sum_{i \in E} \mathbf{P}(i, j) \mathbf{p}(i),$$

that is,
$$\mathbf{p}_n = \mathbf{P}^T \mathbf{p}_{n-1},$$

where $\mathbf{P}^T$ is the transpose of the matrix $\mathbf{P}$, and hence

$$\mathbf{p}_n = \left(\mathbf{P}^T\right)^n \mathbf{p}_0,$$

In particular, if $X_0 = i$, i.e. X_0 has a one-point distribution at point i, then the above formula yields

$$\mathbf{p}_n(j) = \mathbf{P}^n(i, j) \quad \text{for each } n,$$

which establishes the meaning of the coefficients $\mathbf{P}^n(i, j)$ of the n-th power $\mathbf{P}^n$ of the transition matrix $\mathbf{P}$.

Let A be a set described by the random variable $X_0, \ldots X_{n-1}$, i.e. A is of the form

$$A = \bigcup \{X_0 = i_0, \ldots, X_{n-1} = i_{n-1}\},$$

where the union is taken over some set, say $\mathcal{I}$, of indices $i_0, \ldots, i_{n-1}$. Then we have

$$P(X_{n+1} = j | (X_n = i \text{ and } A)) = \mathbf{P}(i, j) \tag{7.7}$$

In fact,

$$P(X_{n+1} = j | (X_n = i \text{ and } A))$$

$$= \frac{P(X_{n+1} = j, X_n = i, A)}{P(X_n = i, A)}$$

$$= \frac{\sum P(X_{n+1} = j, X_n = i, X_{n-1} = i_{n-1}, \ldots X_0 = i_0)}{\sum P(X_n = i, X_{n-1} = i_{n-1}, \ldots X_0 = i_0)},$$

where both sums are taken over the index set $\mathcal{I}$. By Property 2 in the definition of a Markov chain

$$P(X_{n+1} = j, X_n = i, X_{n-1} = i_{n-1}, \ldots X_0 = i_0)$$

$$= P(X_{n+1} = j | (X_n = i, X_{n-1} = i_{n-1}, \ldots X_0 = i_0))$$

$$\times P(X_n = i, X_{n-1} = i_{n-1}, \ldots X_0 = i_0)$$

$$= P(X_{n+1} = j | X_n = i) P(X_n = i, X_{n-1} = i_{n-1}, \ldots X_0 = i_0)$$

$$= \mathbf{P}(i, j) P(X_n = i, X_{n-1} = i_{n-1}, \ldots X_0 = i_0).$$

We then have

$$P(X_{n+1} = j | (X_n = i \text{ and } A)) = \mathbf{P}(i, j),$$

which is the same as (7.7).

The following theorem provides another, more general, interpretation of the coefficients $\mathbf{P}^k(i, j)$ of the matrix $\mathbf{P}^k$ as the probability of passing in k steps from state i to state j.

Theorem 7.2.1 *For each $k \geq 1$, $n \geq 0$ and $i, j \in E$ we have*

$$P(X_{n+k} = j | X_n = i) = \mathbf{P}^k(i, j). \tag{7.8}$$

Proof. For $k = 1$ formula (7.8) follows from Property 2 in the definition of a Markov chain. Assume for an induction proof that (7.8) holds for some k. We will prove it for $k + 1$. Thus we have

$$P(X_{n+k+1} = j | X_n = i) = \frac{P(X_{n+k+1} = j, X_n = i)}{P(X_n = i)}$$

$$= \frac{\sum_{l \in E} P(X_{n+k+1} = j, X_{n+k} = l, X_n = i)}{P(X_n = i)}$$

$$= \frac{\sum_{l \in E} P(X_{n+k+1} = j | X_{n+k} = l, X_n = i) P(X_{n+k} = l, X_n = i)}{P(X_n = i)}.$$

By (7.7) and the induction assumption,

$$P(X_{n+k+1} = j | X_n = i)$$

$$= \frac{\sum_{l \in E} P(X_{n+k+1} = j | X_{n+k} = l) P(X_{n+k} = l | X_n = i) P(X_n = i)}{P(X_n = i)}$$

$$= \sum_{l \in E} \mathbf{P}(l, j) \mathbf{P}^k(i, l) = \mathbf{P}^{k+1}(i, j),$$

which proves the theorem.   □

Now let MAPLE examine some powers of the transition matrix of a random walk.

MAPLE ASSISTANCE 74

```
>  restart:
>  N := 2: A := -N: B := N:
>  q := 0.3: p := 0.5: sa := 0.9: sb := 0.1: r := 1 - p - q:
>  with(linalg):
```

Warning, new definition for norm

Warning, new definition for trace
```
>  dimP := 2*N + 1:
>  P := matrix(dimP,dimP,[0$dimP*dimP]):
>  P[1,1] := sa: P[1,2] := 1 - sa: P[dimP,dimP] := sb:
   P[dimP,dimP-1] := 1 - sb:
   for i from 2 to dimP - 1 do
   P[i,i-1] := q: P[i,i] := r: P[i,i+1] := p
   od:
>  P0 := P:
>  for n from 1 to 10 do P.n := evalm(P.(n-1)&*P0) od:
>  print(P10);
```

$\quad$ [.5448343620, .1153284842, .1217624922, .1449869693,
$\quad$ .07308769250]
$\quad$ [.3459854525, .1028224493, .1592658835, .2516929086,

.1402333063]

[.2191724859, .09555953010, .1807806991, .3246936532,
.1797936318]

[.1565859268, .09060944706, .1948161919, .3533934911,
.2045949431]

[.1420824743, .09087118245, .1941771223, .3682708975,
.2045983235]

We would like to display this output in a more readable form.

```
>  map(x->evalf(x,5),%);
```

$$\begin{bmatrix} .54483 & .11533 & .12176 & .14499 & .073088 \\ .34599 & .10282 & .15927 & .25169 & .14023 \\ .21917 & .095560 & .18078 & .32469 & .17979 \\ .15659 & .090609 & .19482 & .35339 & .20459 \\ .14208 & .090871 & .19418 & .36827 & .20460 \end{bmatrix}$$

Thus, the probability of passing in exactly 10 steps from the left barrier into itself is 0.54483, from the left barrier to the right barrier 0.073088 and from the origin to itself 0.18078.

7.2.3 Irreducible and Transient Chains

From now on will consider only Markov chains for which any two states communicate, i.e. for which the probability $P^k(i,j)$ of passing from any state i to any state j in k steps is positive for some $k = k(i,j)$. Such Markov chains are said to be *irreducible*. Most of the Markov chains that are important in applications are irreducible. However, there are also Markov chains that do not satisfy this property, an example being a random walk with an absorbing barrier, since the probability of passing from the absorbing barrier to any other state is zero.

For an irreducible Markov chain, let $f_n(i)$ denote the *probability of the first return to state i in exactly n steps*, that is

$$f_n(i) = P(X_n = i, X_{n-1} \neq i, \ldots, X_1 \neq i | X_0 = i).$$

Then $F(i)$ defined by

$$F(i) = \sum_{n=1}^{\infty} f_n(i)$$

is the *probability of the first return to state i in a finite time*.

Since $F(i)$ is a probability, it cannot be greater than 1. We will call a state i *recurrent* if $F(i) = 1$ and *transient* if $F(i) < 1$. It can be proved for an irreducible Markov chain that either all states are recurrent or all of them are transient. Thus, we can say that the Markov chain itself is recurrent or transient depending on the nature of its states.

The following theorem, which we stated without proof, allows us to check whether an irreducible Markov chain is recurrent or transient. Denote

$$\mathbf{P}(i) = \sum_{n=1}^{\infty} \mathbf{P}^n(i, i).$$

Theorem 7.2.2 *Let $i \in E$ be a fixed state of an irreducible Markov chain. Then:*

1. *The state i is irreducible if and only if $\mathbf{P}(i) = \infty$.*

2. *If the state i is transient, then $F(i) = \dfrac{\mathbf{P}(i)}{1 + \mathbf{P}(i)}$.*

The numbers $\mathbf{P}(i)$ also have the following interpretation. Define r_i as the total number of returns to the state i.

Theorem 7.2.3 *For each $i \in E$, $\mathbb{E}(r_i) = \mathbf{P}(i)$.*

Proof. Assume that at time 0 the system is at state i. Then $\mathbf{p}(i) = 1$ with $\mathbf{p}(j) = 0$ for $j \neq i$, so

$$P(X_n = i) = P(X_n = i | X_0 = i) = \mathbf{P}^n(i, i)$$

and the expectation of the indicator function $I_{\{X_n = i\}}$ is $\mathbf{P}^n(i, i)$. However, $r_i = \sum_{n=1}^{\infty} I_{\{X_n = i\}}$, from which the result follows. $\square$

Example 7.2.6 Consider a one dimensional random walk with no barriers and $p = q = \dfrac{1}{2}$, as in Example 7.2.1. It is clearly an irreducible Markov chain and $\mathbf{P}^n(i, i) = \mathbf{P}^n(0, 0)$ for any $i \in \mathbb{Z}$, where

$$\mathbf{P}^n(0,0) = \begin{cases} 0 & \text{if } n = 2k - 1 \\ \dfrac{\binom{2k}{k}}{2^{2k}} & \text{if } \quad n = 2k \end{cases}$$

We will sum the $\mathbf{P}^n(0, 0)$ using MAPLE.

MAPLE ASSISTANCE 75
```
>  sum(binomial(2*k,k)/4^k,k=1..infinity);
```
$$\infty$$

Thus, the standard one dimensional random walk is recurrent. It can also be proved that the two dimensional random walk is recurrent, so consider a random walk in at least three dimensional space.

Example 7.2.7 Let $\mathbf{P}_d$ denote the transition matrix of the random walk described in Example 7.2.3 with $d \geq 3$. Fix $i = (i_1, \ldots, i_d) \in \mathbb{Z}^n$. Now passing in n steps from state i into itself in an d-dimensional walk is equivalent to passing from each i_j into itself in n steps in each of the projected one dimensional random walks. By independence we then have

$$\mathbf{P}_d^n(i, i) = \mathbf{P}^n(i_1, i_1) \cdot \ldots \cdot \mathbf{P}^n(i_d, i_d) = (\mathbf{P}^n(0, 0))^d.$$

Let us compute $\mathbf{P}_d^n(i)$ for $d = 3, 4, 5, 6$ using MAPLE.

MAPLE ASSISTANCE 76

```
> seq(sum((binomial(2*k,k)/4^k)^d,
    k = 1..infinity), d = [3,4,5,6]);
```

$$\frac{1}{8} \, \text{hypergeom}([1, \frac{3}{2}, \frac{3}{2}, \frac{3}{2}], [2, 2, 2], 1),$$

$$\frac{1}{16} \, \text{hypergeom}([1, \frac{3}{2}, \frac{3}{2}, \frac{3}{2}, \frac{3}{2}], [2, 2, 2, 2], 1),$$

$$\frac{1}{32} \, \text{hypergeom}([1, \frac{3}{2}, \frac{3}{2}, \frac{3}{2}, \frac{3}{2}, \frac{3}{2}], [2, 2, 2, 2, 2], 1), \frac{1}{64}$$

$$\text{hypergeom}([1, \frac{3}{2}, \frac{3}{2}, \frac{3}{2}, \frac{3}{2}, \frac{3}{2}, \frac{3}{2}], [2, 2, 2, 2, 2, 2], 1)$$

```
> evalf([%]);
```

$$[.3932039296, .1186363871, .04682554981, .02046077706$$
$$]$$

Compute corresponding probabilities $F_d(0)$.

```
> map(x ->x/(1+x),%);
```

$$[.2822299888, .1060544682, .04473099631, .02005052768$$
$$]$$

A d-dimensional random walk with $d > 2$ is thus transient, with the mean number of returns to a given state and the probability of return decreases with increasing dimension.

Consider an irreducible Markov chain and fix some state $i \in E$. Since i communicates with itself there is a number $n \geq 1$ such that $\mathbf{P}^n(i, i) > 0$. Let N_i be the set of all such n's and note that this set has the property that if $m, n \in N_i$ then $m + n \in N_i$, which follows from more general observations that

$$\mathbf{P}^{m+n}(i, j) = \sum_{l \in E} \mathbf{P}^m(i, l) \mathbf{P}^n(l, j) \geq \mathbf{P}^m(i, k) \mathbf{P}^n(k, j)$$

for all states i, j and k, In particular, $\mathbf{P}^{m+n}(i,i) \geq \mathbf{P}^m(i,i)\mathbf{P}^n(i,i) > 0$.

We will say that state i is *periodic with period* $\nu > 1$ if ν is the greatest common divisor of the set N_i. One can prove for an irreducible Markov chain that either all states are periodic with the same period or none of them are periodic. In the first situation we will say that the Markov chain is periodic with an appropriate period.

The random walk described in Example 7.2.1 is obviously periodic with period 2. On the other hand, a random walk with $p + q < 1$ and without barriers is not periodic. The existence non-absorbing barriers may also result in non-periodicity, even when $p + q = 1$.

Often we may be interested in the long time behavior of a Markov chain, in particular in the asymptotic distribution of X_n's, if it exists. We now consider the so called Ergodic Theorem, one of the most powerful theorems in the theory of dynamical systems, in a very simple situation.

Theorem 7.2.4 *Consider an irreducible Markov chain with finite set E of the states with k denoting the number of states. Then, either*

1. *The chain is periodic,*

   or

2. *There exists a vector π with components $\pi_1, \ldots, \pi_k$ such that*
   a) $\pi_i > 0$ for each $i \in E$;
   b) For each $i, j \in E$,

$$\lim_{n \to \infty} \mathbf{P}^n(i,j) = \pi_j.$$

   Then,
   c) The vector π is the unique solution of the equation

$$\mathbf{P}^T x = x$$

   satisfying the condition $\sum_{i \in E} x_i = 1$.

In second case, we say that the chain is ergodic and call the vector π a stationary distribution.

The ergodicity of the Markov chain means that for large n the probability of passing from state i to state j in n steps is positive and does not depend on the first state i but rather on the final state j. Statement (c) means that all these probabilities can be found by solving the appropriate system of linear equations.

We will not present the advanced proof of this theorem, but instead we shall check the validity of the theorem experimentally. Consider, for example, the random walk with two barriers and the parameters defined on page 204. We have already found the matrix $\mathbf{P}^{10}$, so we will now use MAPLE to compute some higher powers of $\mathbf{P}$ and then compare them with the theoretical result of the above theorem.

MAPLE ASSISTANCE 77

```
>  P0 := P:
>  for n from 1 to 50 do P.n := evalm(P.(n-1)&*P0) od:
>  map(x->evalf(x,5),P30);
```

$$\begin{bmatrix} .34518 & .10291 & .15862 & .25369 & .13959 \\ .30874 & .10064 & .16533 & .27356 & .15173 \\ .28551 & .099199 & .16961 & .28622 & .15946 \\ .27399 & .098482 & .17173 & .29250 & .16329 \\ .27137 & .098319 & .17222 & .29393 & .16417 \end{bmatrix}$$

```
>  map(x->evalf(x,5),P50);
```

$$\begin{bmatrix} .30859 & .10063 & .16536 & .27364 & .15178 \\ .30190 & .10022 & .16659 & .27729 & .15400 \\ .29765 & .099954 & .16737 & .27961 & .15542 \\ .29553 & .099823 & .16776 & .28076 & .15612 \\ .29505 & .099793 & .16785 & .28102 & .15628 \end{bmatrix}$$

The rows of successive powers of **P** should approach the transposition of the stationary distribution π promised by the theorem. We will find this distribution by solving the appropriate system of linear equations.

If we want to use a MAPLE solver, we first have to determine an appropriate system of linear equations of $Ax = b$ which corresponds to the conditions in statement (c). We can use MAPLE to determine this matrix A and vector b.

```
>  J := diag(1$dimP):
>  d := matrix(dimP,1,[1$dimP]):
   b := matrix(dimP+1,1,[0$dimP,1]):
>  A := transpose(augment(P-J,d)):
```

We solve the system.

```
>  linsolve(A,b);
```

$$\begin{bmatrix} .3003708282 \\ .1001236094 \\ .1668726823 \\ .2781211372 \\ .1545117429 \end{bmatrix}$$

```
>  map(x->evalf(x,5),transpose(%));
```

$$\begin{bmatrix} .30037 & .10012 & .16687 & .27812 & .15451 \end{bmatrix}$$

7.3 Special Classes of Stochastic Processes

The Markov chains discussed in previous Section are an example of stochastic processes, i.e. a family of random variables or vectors depending on time. In general, we will denote the time set under consideration by $\mathbb{T}$ and assume that there is a common underlying probability space $(\Omega, \mathcal{A}, P)$. A *stochastic process* $X = \{X(t),\ t \in \mathbb{T}\}$ is thus a function of two variables $X : \mathbb{T} \times \Omega \longrightarrow \mathbb{R}^d$ where $X(t) = X(t, \cdot)$ is a random vector for each $t \in \mathbb{T}$. For each $\omega \in \Omega$ we call $X(\cdot, \omega) : \mathbb{T} \longrightarrow \mathbb{R}^d$ a *realization*, a *sample path* or a *trajectory* of the stochastic process.

Markov chains are also an example of *discrete-time stochastic processes*, i.e. processes with $\mathbb{T}$ discrete, most often $\mathbb{T} = \{0, 1, 2, \ldots\}$. If $\mathbb{T}$ is an interval, typically $[0, \infty)$, then we have a *continuous-time stochastic process*.

We will sometimes write X_t instead of $X(t)$ and P_t instead of P_{X_t}, where as usual P_{X_t} is the probability distribution of the random variable or vector X_t. A probability distribution of the kd-dimensional random vector $(X_{t_1}, \ldots, X_{t_k})$ will be called a *finite joint probability distribution*, joint distribution for short, and denoted by $P_{t_1, \ldots, t_k}$. The collection of all finite joint probability distributions is called the *probability law* of the process.

A sequence of *independent and identically distributed* (in brief, i.i.d.) random variables is a trivial example of a discrete time stochastic process in the sense that what happens at one instant is completely unaffected by what happens at any other, past or future, instant. Moreover, its probability law is quite simple, since any joint distribution is given by

$$P_{t_1 t_2, \cdots, t_k} = P_{t_1} \times P_{t_2} \times \cdots \times P_{t_k}, \qquad (7.9)$$

where the P_{t_i} are all equal to each other.

Another simple example is given by the iterates of a first order difference equation

$$X_{n+1} = f(X_n) \qquad (7.10)$$

for $n = 0, 1, 2, 3, \ldots$, where f is a given function and X_0 is a random variable. Here each random variable X_n is determined exactly from its predecessor, and ultimately from the initial random variable X_0. Pseudo-random numbers are an example of this kind of process.

The most interesting stochastic processes encountered in applications possess a relationship between random vectors at different time instants that lies somewhere between the two extremes above. Markov chains with their laws being uniquely determined by the initial distribution $\mathbf{p}$ and the transition matrix $\mathbf{P}$ are a good example of such a situation.

Another example is a *Gaussian process*, i.e. a stochastic process (one dimensional here for simplicity) having all its joint probability distributions Gaussian. It may also satisfy (7.9), but generally need not. If $X(t)$ is a Gaussian process, then $X(t) \sim N\big(\mu(t), \sigma^2(t)\big)$ for all $t \in \mathbb{T}$, where μ and σ are some functions.

For both continuous and discrete time sets $\mathbb{T}$, it is useful to distinguish various classes of stochastic processes according to their specific temporal relationships.

An important class of stochastic process are those with *independent increments*, that is for which the random variables $X(t_{j+1}) - X(t_j)$, $j = 0, 1, 2, \ldots, n - 1$ are independent for any finite combination of time instants $t_0 < t_1 < \cdots < t_n$ in $\mathbb{T}$. If t_0 is the smallest time instant in $\mathbb{T}$, then the random variables $X(t_0)$ and $X(t_j) - X(t_0)$ for any other t_j in $\mathbb{T}$ are also required to be independent. Markov chains are processes with independent increments.

The *stationary processes* are another interesting class of stochastic processes since they represent a form of probabilistic equilibrium in the sense that the particular instants at which they are examined are not important. We say that a process is *strictly stationary* if its joint distributions are all invariant under time displacements, that is if

$$P_{t_1+h,t_2+h,\cdots,t_n+h} = P_{t_1,t_2,\cdots,t_n} \tag{7.11}$$

for all t_i, $t_i + h \in \mathbb{T}$ where $i = 1, 2, \ldots, n$ and $n = 1, 2, 3, \ldots$. In particular, the $X(t)$ have the same distribution for all $t \in \mathbb{T}$.

For example, a sequence of i.i.d. random variables is strictly stationary. On the other hand, a Markov chain is strictly stationary if and only if its initial distribution $\mathbf{p_0}$ is a stationary distribution, see page 208.

In most cases we understand stationarity of a stochastic process in a weaker form. We define it here for one dimensional processes, that is for which the $X(t)$ are random variables. Assuming they exist, the expectations and variances

$$\mu(t) = \mathbb{E}(X(t)), \quad \sigma^2(t) = \mathrm{Var}\,(X(t)) \tag{7.12}$$

at each instant $t \in \mathbb{T}$ and the covariances

$$C(s,t) = \mathbb{E}((X(s) - \mu(s))(X(t) - \mu(t))) \tag{7.13}$$

at distinct instants s, $t \in \mathbb{T}$ provide some information about the time variability of a stochastic process. In particular, if there is a constant μ and a function $c : \mathbb{R} \longrightarrow \mathbb{R}^+$ such that the means, variances and covariances of a stochastic process satisfy

$$\mu(t) = \mu, \quad \sigma^2(t) = c(0) \quad \text{and} \quad C(s,t) = c(t - s) \tag{7.14}$$

for all s, $t \in \mathbb{T}$, then we call the process *wide-sense stationary*. This means that the process is stationary only with respect to its first and second moments.

7.4 Continuous-Time Stochastic Processes

Markov chains are examples of discrete-time stochastic processes, whereas the simple Poisson process mentioned on page 135 is an example of a continuous time process.

Definition 7.4.1 *A* Poisson process *with intensity parameter* $\lambda > 0$ *is a stochastic process* $X = \{X(t), t \geq 0\}$ *such that*

1. $X(0) = 0$;
2. $X(t) - X(s)$ *is a Poisson distributed random variable with parameter* $\lambda(t - s)$ *for all* $0 \leq s < t$;
3. *the increments* $X(t_2) - X(t_1)$ *and* $X(t_4) - X(t_3)$ *are independent for all* $0 < t_1 < t_2 \leq t_3 < t_4$.

One can show that the means, variances and covariances of the Poisson process are

$$\mu(t) = \lambda t, \qquad \sigma^2(t) = \lambda t, \qquad C(s,t) = \lambda \min\{s,t\}$$

for all s, $t > 0$.

A Poisson process is an example of a non-stationary stochastic process with independent increments.

Definition 7.4.2 *The* Ornstein-Uhlenbeck process $X = \{X(t), t \geq 0\}$ *with parameter* $\gamma > 0$ *and initial value* $X_0 \in N(0,1)$ *is a Gaussian process with means and covariances*

$$\mu(t) = 0, \qquad C(s,t) = e^{-\gamma|t-s|}$$

for all s, $t \in \mathbb{R}^+$.

The Ornstein-Uhlenbeck process is clearly a wide-sense stationary process. It is also strictly stationary.

7.4.1 Wiener Process

There is a very fundamental stochastic process, now called a *Wiener process*, that was proposed by Norbert Wiener as a mathematical description of physical process of Brownian motion, the erratic motion of a grain of pollen on a water surface due to continual bombardment by water molecules. The Wiener process is sometimes called *Brownian motion*, but we will use separate terminology here to distinguish between the mathematical and physical processes.

Definition 7.4.3 *The standard* Wiener process $W = \{W(t), t \geq 0\}$ *is a Gaussian process with independent increments for which*

$$W(0) = 0, \ w.p.1, \quad \mathbb{E}(W(t)) = 0, \quad \text{Var}(W(t) - W(s)) = t - s \quad (7.15)$$

for all $0 \leq s \leq t$.

A Wiener process is *sample-path continuous*, that is, its sample paths are, almost surely, continuous functions of time. We see from (7.15) that the variance grows without bound as time increases while the mean always remains zero, which means that many sample paths must attain larger and larger values, both positive and negative, as time increases. A Wiener process is, however, not differentiable at any $t \geq 0$. This is easy to see in the mean-square sense, since

$$\mathbb{E}\left(\left(\frac{W(t+h) - W(t)}{h}\right)^2\right) = \frac{\mathbb{E}\left((W(t+h) - W(t))^2\right)}{h^2} = \frac{1}{h}.$$

It can also be shown, but with considerably more difficulty, that the sample paths of a Wiener process are, almost surely, nowhere differentiable functions of time.

By definition the initial value of a Wiener process is $W(0) = 0$, w.p.1. In some applications, however, it is useful to have a modification of a Wiener process for which the sample paths all pass through the same initial point x, not necessarily 0, and a given point y at a later time $t = T$. Such a process, denoted by $B_{0,x}^{T,y}(t)$, is defined sample pathwise for $0 \leq t \leq T$ by

$$B_{0,x}^{T,y}(t,\omega) = x + W(t,\omega) - \frac{t}{T}\{W(T,\omega) - y + x\} \qquad (7.16)$$

and is called a *Brownian bridge* or a *tied-down Wiener process*. It is a Gaussian process satisfying the constraints $B_{0,x}^{T,y}(0,\omega) = x$ and $B_{0,x}^{T,y}(T,\omega) = y$, so in a manner of speaking can be considered to be a kind of conditional Wiener process. Since it is Gaussian it is determined uniquely by its means and covariances, which are

$$\mu(t) = x - \frac{t}{T}(x - y) \quad \text{and} \quad C(s,t) = \min\{s,t\} - \frac{st}{T} \qquad (7.17)$$

for $0 \leq s, t \leq T$.

We can approximate a standard Wiener process in distribution on any finite time interval by a scaled random walk. For example, we can subdivide a given interval $[0, T]$ into N subintervals

$$0 = t_0^{(N)} < t_1^{(N)} < \cdots < t_k^{(N)} < \cdots < t_N^{(N)} = T$$

of equal length $\Delta t = T/N$ and construct a stepwise continuous random walk $S_N(t)$ by taking independent, equally probable steps of length $\pm\sqrt{\Delta t}$ at the end of each subinterval. Let $X_1, \ldots, X_N$ be an i.i.d. sequence of random variables $X_1, \ldots, X_N, \ldots$ with mean $= 0$ and variance $= 1$. Then we define

$$S_N(t_n^{(N)}) = (X_1 + X_2 + \cdots + X_n)\sqrt{\Delta t} \qquad (7.18)$$

with

$$S_N(t) = S_N(t_n^{(N)}) \qquad (7.19)$$

on $t_n^{(N)} \leq t < t_{n+1}^{(N)}$ for $n = 0, 1, \ldots, N-1$, where $S_N(0) = 0$. This random walk has independent increments $X_1\sqrt{\Delta t}, X_2\sqrt{\Delta t}, X_3\sqrt{\Delta t}, \ldots$ for the given subintervals, but is not a process with independent increments. It can be shown that

$$\mathbb{E}(S_N(t)) = 0, \quad \text{Var}(S_N(t)) = \Delta t \left[\frac{t}{\Delta t}\right]$$

for $0 \leq t \leq T$, where $[\tau]$ denotes the integer part of $\tau \in \mathbb{R}$, i.e. $[\tau] = k$ if $k \leq \tau < k+1$ for some integer k. Now, $\text{Var}(S_N(t)) \longrightarrow t$ as $N = 1/\Delta t \longrightarrow \infty$ for any $0 \leq t \leq T$. Similarly, it can be shown for any $0 \leq s < t \leq T$ that $\text{Var}(S_N(t) - S_N(s)) \longrightarrow t - s$ as $N \longrightarrow \infty$. Hence it follows by the Central Limit Theorem that $S_N(t)$ converges in distribution as $N \longrightarrow \infty$ to a process with independent increments satisfying conditions (8.1), i.e. to a standard Wiener process.

Example 7.4.1 Consider independent random variables X_i taking values ± 1 with probability $\frac{1}{2}$ each and define

$$Y_1^{(k)} = (X_1 \ldots + X_k)/\sqrt{k}$$
$$Y_2^{(k)} = (X_{k+1} + \ldots + X_{2k})/\sqrt{k}$$

$$\ldots\ldots\ldots$$

for $k = 1, 2, 3, \ldots$. Thus, the $Y_n^{(k)}$ also form a sequence of i.i.d. random variables with zero mean and the unit variance. Note that large k means possibly long steps during the walk, whereas small k, in particular $k = 1$, means very short steps. This random walk, which is essentially the same as the one above, is a better approximation of a Wiener process.

MAPLE ASSISTANCE 78

We will simulate the above situation.
```
> with(stats):
```

Write a procedure that can plot approximation of the Wiener process given the steps and the lengths of the time interval.
```
> Wpath := proc(steps,t)
> local walk, i, N,ww: N := nops(steps):
> walk[0] := 0:
> for i from 0 to N-1 do
    walk[i+1] := walk[i] + steps[i+1]*sqrt(t/N)
    od:
    ww := seq(plot(walk[i],t*i/N..t*(i+1)/N), i = 0..N - 1):
> plots[display]([ww]):
    end:
```

Choose $N = 400$ of $\pm$ steps according to the uniform distribution and arrange them into six groups with $k = 40, 20, 10, 5, 2, 1$.

```
>  N := 400:
>  numbers := [random[empirical[0.5,0.5]](N)]:
   st1 := map(x -> 2*x - 3, numbers):
>  list_of_k := [40,20,10,5,2,1]:
>  for j from 1 to nops(list_of_k) do
   k := list_of_k[j]:
   st[k] := [seq(sum(st1[p],p = k*i - k + 1..k*i)/sqrt(k),
   i = 1..N/k)]
   od:
```

Display the results (use the MAPLE animation facility if possible).

```
>  plots[display]([seq(Wpath(st[k],4),k = list_of_k)],
   insequence = true,axes = FRAMED, tickmarks = [2,3]):
>  plots[display](%);
```

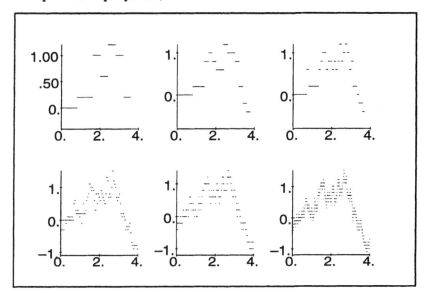

7.4.2 Markov Processes

Markov processes to be defined here are an extension of Markov chains to the continuous-time case. The space of states need not be finite or countable. For simplicity we consider here only one dimensional Markov processes, the generalization to higher dimensions being straightforward.

Suppose that every joint distribution $P_{t_1,t_2,\ldots,t_k}$ for $k = 1, 2, 3, \ldots$ of a continuous time stochastic process process $X = \{X(t),\, t \in \mathbb{R}^+\}$ is discrete and supported at points x_j or has a density $\mathbf{p} = \mathbf{p}(t_1, x_1; t_2, x_2; \ldots; t_k, x_k)$. Consider its conditional probabilities

$$P\left(X(t_{n+1}) \in B \mid X(t_1) = x_1, X(t_2) = x_2, \ldots, X(t_n) = x_n\right)$$

for all Borel subsets B of $\mathbb{R}$, time instants $0 < t_1 < t_2 < \ldots < t_n < t_{n+1}$ and all states $x_1, x_2, \ldots, x_n \in \mathbb{R}$, see Example 7.1.4.

Definition 7.4.4 *A continuous-time stochastic process* $X = \{X(t), t \geq 0\}$ *is called a* Markov process *if it satisfies the* Markov property, *i.e.*

$$P\left(X(t_{n+1}) \in B \mid X(t_1) = x_1, X(t_2) = x_2, \ldots, X(t_n) = x_n\right) \quad (7.20)$$

$$= P\left(X(t_{n+1}) \in B \mid X(t_n) = x_n\right)$$

for all Borel subsets B of $\mathbb{R}$, all time instants $0 < t_1 < t_2 < \ldots < t_n < t_{n+1}$ and all states $x_1, x_2, \ldots, x_n \in \mathbb{R}$ for which the conditional probabilities are defined.

Let X be a Markov process and write its *transition probabilities* as

$$\mathbf{P}(s, x; t, B) = P(X(t) \in B \mid X(s) = x), \qquad 0 \leq s < t.$$

If the probability distribution P is discrete, the transition probabilities are uniquely determined by the transition matrix with components

$$\mathbf{P}(s, i; t, j) = P(X(t) = x_j \mid X(s) = x_i)$$

depending also on s and t, whereas in the continuous state case

$$\mathbf{P}(s, x; t, B) = \int_B \mathbf{p}(s, x; t, y)\, dy$$

for all $B \in \mathcal{B}$, where the density $p(s, x; t, \cdot)$ is called the *transition density*.

An example of a Markov process with a finite space of states is *telegraphic noise*. This is a continuous-time Markov process $\{X(t), t \geq 0\}$ taking values -1 and 1 only with the transition probabilities

$$P(X(t) = 1 \mid X(s) = 1) = P(X(t) = -1 \mid X(s) = -1) = \frac{1 + e^{-(t-s)}}{2}$$
$$P(X(t) = 1 \mid X(s) = -1) = P(X(t) = -1 \mid X(s) = 1) = \frac{1 - e^{-(t-s)}}{2}$$

for all $0 \leq s < t$, so its transition matrix is

$$\begin{bmatrix} \dfrac{1 + e^{-(t-s)}}{2} & \dfrac{1 - e^{-(t-s)}}{2} \\ \dfrac{1 - e^{-(t-s)}}{2} & \dfrac{1 + e^{-(t-s)}}{2} \end{bmatrix}$$

A Markov process is said to be *homogeneous* if all of its transition probabilities depend only on the time difference $t - s$ rather than on the specific values of s and t. The telegraphic noise process is a homogeneous Markov processes as are the standard Wiener process and the Ornstein-Uhlenbeck process with parameter $\gamma = 1$, for which the transition densities are, respectively,

$$\mathbf{p}(s,x;t,y) = \frac{1}{\sqrt{2\pi(t-s)}} \exp\left(-\frac{(y-x)^2}{2(t-s)}\right) \tag{7.21}$$

and

$$\mathbf{p}(s,x;t,y) = \frac{1}{\sqrt{2\pi\left(1-e^{-2(t-s)}\right)}} \exp\left(-\frac{\left(y-xe^{-(t-s)}\right)^2}{2\left(1-e^{-2(t-s)}\right)}\right). \tag{7.22}$$

Since $\mathbf{p}(s,x;t,y) = \mathbf{p}(0,x;t-s,y)$ we usually omit the superfluous first variable and simply write $\mathbf{p}(x;t-s,y)$, with $\mathbf{P}(x;t-s,B) = \mathbf{P}(0,x;t-s,B)$ for such transition probabilities.

The transition density of the standard Wiener process is obviously a smooth function of its variables for $t > s$. Evaluating the partial derivatives of (7.21) we find that they satisfy the partial differential equations

$$\frac{\partial p}{\partial t} - \frac{1}{2}\frac{\partial^2 p}{\partial y^2} = 0, \qquad (s,x) \text{ fixed}, \tag{7.23}$$

and

$$\frac{\partial p}{\partial s} + \frac{1}{2}\frac{\partial^2 p}{\partial x^2} = 0, \qquad (t,y) \text{ fixed}. \tag{7.24}$$

The first equation is an example of a heat equation which describes the variation in temperature as heat diffuses through a physical medium. The same equation describes the diffusion of a chemical substance such as an inkspot. It should thus not be surprising that a standard Wiener process serves as a prototypical example of a (stochastic) diffusion process.

7.4.3 Diffusion Processes

Diffusion processes, which we now define in the one dimensional case, are a rich and useful class of Markov processes.

Definition 7.4.5 *A Markov process with transition densities* $\mathbf{p}(s,x;t,y)$ *is called a* diffusion process *if the following three limits exist for all* $\epsilon > 0$, $s \geq 0$ *and* $x \in \mathbb{R}$.

$$\lim_{t \downarrow s} \frac{1}{t-s} \int_{|y-x|>\epsilon} \mathbf{p}(s,x;t,y)\, dy = 0, \tag{7.25}$$

$$\lim_{t \downarrow s} \frac{1}{t-s} \int_{|y-x|<\epsilon} (y-x)\mathbf{p}(s,x;t,y)\, dy = a(s,x) \tag{7.26}$$

and

$$\lim_{t \downarrow s} \frac{1}{t-s} \int_{|y-x|<\epsilon} (y-x)^2\mathbf{p}(s,x;t,y)\, dy = b^2(s,x), \tag{7.27}$$

where a *and* b *are well-defined functions.*

Condition (7.25) prevents a diffusion process from having instantaneous jumps. The quantity $a(s, x)$ is called the *drift* of the diffusion process and $b(s, x)$ its *diffusion coefficient* at time s and position x. (7.26) implies that

$$a(s, x) = \lim_{t \downarrow s} \frac{1}{t - s} \mathbb{E}\left(X(t) - X(s) | X(s) = x\right), \qquad (7.28)$$

so the drift $a(s, x)$ is the instantaneous rate of change in the mean of the process given that $X(s) = x$. Similarly, it follows from (7.27) that the squared diffusion coefficient

$$b^2(s, x) = \lim_{t \downarrow s} \frac{1}{t - s} \mathbb{E}\left((X(t) - X(s))^2 | X(s) = x\right) \qquad (7.29)$$

denotes the instantaneous rate of change of the squared fluctuations of the process given that $X(s) = x$.

When the drift a and diffusion coefficient b of a diffusion process are sufficiently smooth functions, the transition density $\mathbf{p}(s, x; t, y)$ also satisfies partial differential equations, which reduce to (7.23) and (7.24) for a standard Wiener process. These are the *Kolmogorov forward equation*

$$\frac{\partial p}{\partial t} + \frac{\partial}{\partial y}\left\{a(t, y)p\right\} - \frac{1}{2}\frac{\partial^2}{\partial y^2}\left\{b^2(t, y)p\right\} = 0, \qquad (s, x) \text{ fixed}, \qquad (7.30)$$

and the *Kolmogorov backward equation*

$$\frac{\partial p}{\partial s} + a(s, x)\frac{\partial p}{\partial x} + \frac{1}{2}b^2(s, x)\frac{\partial^2 p}{\partial x^2} = 0, \qquad (t, y) \text{ fixed}, \qquad (7.31)$$

with the former giving the forward evolution with respect to the final state (t, y) and the latter giving the backward evolution with respect to the initial state (s, x). The forward equation (7.30) is commonly called the *Fokker-Planck equation*, especially by physicists and engineers.

We would expect that a diffusion process should have well-behaved sample paths. In fact these are, almost surely, continuous functions of time, although they need not be differentiable, as we have already seen for Wiener processes.

7.5 Continuity and Convergence

We can define the continuity of a continuous-time stochastic process $X(t)$ in t in several different ways corresponding to the different types of convergences of the sequences of random variables that were introduced in Section 4.5. Suppose that the process has the underlying probability space $(\Omega, \mathcal{A}, \mathbb{P})$ and let t be fixed .

I. *Continuity with probability one:*

$$\mathbb{P}\left(\{\omega \in \Omega : \lim_{s \to t} |X(s, \omega) - X(t, \omega)| = 0\}\right) = 1; \qquad (7.32)$$

II. *Mean-square continuity:* $\mathbb{E}(X(t)^2) < \infty$ and

$$\lim_{s \to t} \mathbb{E}\left(|X(s) - X(t)|^2\right) = 0; \tag{7.33}$$

III. *Continuity in probability* or *stochastic continuity:*

$$\lim_{s \to t} \mathbb{P}(\{\omega \in \Omega : |X(s, \omega) - X(t, \omega)| \geq \epsilon\}) = 0 \text{ for all } \epsilon > 0; \tag{7.34}$$

IV. *Continuity in distribution:*

$$\lim_{s \to t} F_s(x) = F_t(x) \quad \text{for all continuity points of } F_t. \tag{7.35}$$

These are related to each other in the same way as the convergences of random sequences. In particular, **(I)** and **(II)** both imply **(III)**, but not each other, and **(III)** implies **(IV)**. Moreover **(III)** can be written in terms of some metric, namely:

$$\lim_{s \to t} \mathbb{E}\left(\frac{|X(s) - X(t)|}{1 + |X(s) - X(t)|}\right) = 0. \tag{7.36}$$

The random telegraphic noise process defined on page 216 is a continuous-time Markov process with two states -1 and $+1$. Its sample paths are piecewise constant functions jumping between the two values ± 1 with equal probability. It follows from its transition matrices that the covariances are

$$\mathbb{E}\left(|X(s) - X(t)|^2\right) = 2\left(1 - \exp(-|s - t|)\right) \tag{7.37}$$

for all s, $t \geq 0$, so this process is mean-square continuous at any instant t and hence also continuous in probability as well as in distribution. It is also known that the telegraphic noise process is continuous with probability one at any instant t. This may seem surprising in view of the fact that its sample paths are actually discontinuous at the times when a jump occurs. There is, however, a simple explanation. Continuity with probability one at the time instant t means that $\mathbb{P}(A_t) = 0$, where $A_t = \{\omega \in \Omega : \lim_{s \to t} |X(s, \omega) - X(t, \omega)| \neq 0\}$. Continuity of almost all of the sample paths, which is called *sample-path continuity*, requires $\mathbb{P}(A) = 0$ where $A = \cup_{t \geq 0} A_t$. Since A is the uncountable union of events A_t it need not be an event, in which case $\mathbb{P}(A)$ is not defined, or when it is an event it could have $\mathbb{P}(A) > 0$ even though $\mathbb{P}(A_t) = 0$ for every $t \geq 0$. The second of these situations is what happens for the telegraphic noise process.

For a diffusion process the appropriate continuity concept is sample-path continuity. There is a criterion due to Kolmogorov which implies the sample-path continuity of a continuous-time stochastic process $X = \{X(t), t \in \mathbb{T}\}$, namely the existence of positive constants α, β, C and h such that

$$\mathbb{E}(|X(t) - X(s)|^\alpha) \leq C|t - s|^{1+\beta} \tag{7.38}$$

for all s, $t \in \mathbb{T}$ with $|t - s| \leq h$. For example, this is satisfied by the standard Wiener process $W = \{W(t), t \geq 0\}$, for which it can be shown that

$$\mathbb{E}\left(|W(t) - W(s)|^4\right) = 3|t - s|^2 \tag{7.39}$$

for all s, $t \geq 0$. Hence the Wiener process has, almost surely, continuous sample paths.

We conclude this section with the remark that the uncountability of the time set is responsible for many subtle problems concerning continuous time stochastic processes. These require the rigorous framework of measure and integration theory for their clarification and resolution.

7.6 MAPLE Session

M 7.1 *Adjust the procedure* RandWalk *to animate the random walk of a particle along the line.*

We will use the geometry package to get a solid disc.

```
>  with(geometry):
   _EnvHorizontalName := X: _EnvVerticalName := Y:
```

Rewrite the RandWalk procedure.

```
>  RandW := proc(N,p,q,sa,sb,M)
>  local A, B, r, PA, PC, PB, state, step, path,rad,i:
>  A := -N: B := N: r := 1 - p - q:
>  PA := empirical[0.,sa,1-sa]:
   PC := empirical[q,r,p]:
   PB := empirical[1-sb,sb,0.]:
>  state := 0: path := state:
>  from 1 to M do
   if (A < state and state < B) then
   step := trunc(stats[random,PC](1))-2
   elif state = A then step := trunc(stats[random,PA](1))-2
   elif state = B then step := trunc(stats[random,PB](1))-2
   fi:
   state := state + step:
   path := path,state
   od:
>  rad := N/10:
>  seq(draw(circle(dd,[point(S,i,0),rad]), filled = true,
   view = [-N..N,-N..N],axes = NORMAL,ytickmarks = 0),
   i = [path]):
>  plots[display](%,insequence = true):
   end:
```

Prepare the plot for animation.

```
>  RandW(10,0.15,0.25,0.1,0.8,300);
```

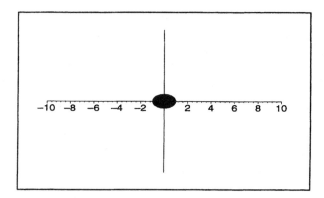

Now, use the context menu to view the animation.

M 7.2 *Write* MAPLE *commands that allow the simulation of the behavior of any Markov chain with a given transition matrix* **P** *and a given initial distribution vector* **p**.

Consider three states only. Input the rows of the transition matrix and the initial distribution. Note that the second state will be hard to reach in this example.

```
> P1 := empirical[0.1,0.,0.9]:
  P2 := empirical[0.2,0.1,0.7]:
  P3 := empirical[0.4,0.1,0.5]:
  init_distr := empirical[0.1,0.8,0.1]:
> T1 := proc() trunc(stats[random,P1](1)) end:
> T2 := proc() trunc(stats[random,P2](1)) end:
> T3 := proc() trunc(stats[random,P3](1)) end:
```

Test the procedures.

```
> T1(),T2(),T3();
```

$$3, 2, 3$$

Initialize the process.

```
> state := trunc(stats[random,init_distr](1));
```

$$state := 2$$

Run the process.

```
> path := state:
  from 1 to 150 do
  state := T.state():
  path := path,state
  od:
  path;
```

2, 1, 3, 1, 3, 2, 3, 1, 3, 1, 3, 3, 3, 3, 3, 1, 3, 1, 3, 2, 3, 1, 3, 2, 1, 3, 1, 3, 1,
3, 3, 1, 3, 3, 3, 3, 1, 3, 1, 3, 3, 1, 3, 3, 3, 3, 1, 1, 3, 1, 3, 3, 1, 3, 1, 3, 1,
3, 3, 1, 3, 1, 1, 3, 3, 1, 3, 1, 3, 2, 3, 2, 3, 1, 3, 3, 3, 1, 1, 1, 3, 3, 3, 1,
3, 3, 3, 2, 3, 1, 3, 3, 1, 3, 3, 3, 3, 3, 2, 3, 2, 3, 1, 3, 3, 3, 3, 1, 3, 3, 1,
3, 3, 2, 3, 1, 1, 1, 3, 3, 3, 1, 3, 1, 3, 3, 3, 3, 1, 3, 1, 3, 1, 3, 3, 3, 1, 3,
1, 1, 3, 2, 1, 3, 3, 1, 3, 3, 3, 1, 3

M 7.3 *Find the distributions of the random variables X_n describing Alice's financial capital after the n-th game in the sequence of games described in Example 7.2.4. Consider the parameters $A = 8$, $B = 5$, $p = 0.2$, $q = 0.4$ corresponding to a situation in which Alice has a larger initial capital, but Barbara is the better player. Inspect these distributions for particular values of n and compute the mathematical expectations of X_n for these n's.*

We modify the commands used for the investigation of the transition matrix for a random walk.

```
> restart:
> with(linalg):
```

Warning, new definition for norm

Warning, new definition for trace

```
> A := 8: B := 5:
> q := 0.4: p := 0.2: r := 1 - p - q:
> dimP := A + B + 1:
> P := matrix(dimP,dimP,[0$dimP*dimP]):
```

Remember that the actual states of our chain are $0, 1, \ldots, A + B$, while the transition matrix is indexed here with $1, 2, \ldots, A + B + 1$.

```
> P[1,1] := 1: P[1,2] := 0:
  P[dimP,dimP] := 1: P[dimP,dimP-1] := 0:
  for i from 2 to dimP - 1 do
  P[i,i-1] := q: P[i,i] := r: P[i,i+1] := p
  od:
> p0 := matrix(dimP,1,[0$dimP]):
> p0[A+1,1] := 1:
> PT := transpose(P):
> for n from 1 to 200 do
  p.n := evalm(PT&*p.(n-1))
  od:
```

Thus we have defined distributions of the Alice's capital after completing each of 200 games. We can look for particular games computing the expectations.

```
> map(x->evalf(x,3),transpose(p5));
```

$$[0, 0, 0, .0102, .0512, .128, .205, .230, .189, .115, .0512,$$
$$.0160, .00320, .00032]$$

```
>  add(p.5[i,1]*(i-1) ,i = 1..dimP);
                    7.00000
>  map(x->evalf(x,3),transpose(p20));
```

$$[.176, .0620, .0954, .113, .118, .112, .0973, .0776, .0568,$$
$$.0378, .0225, .0114, .00422, .0165]$$

```
>  add(p.20[i,1]*(i-1) ,i = 1..dimP);
                 4.157931287
>  map(x->evalf(x,3),transpose(p200));
```

$$[.969, .0000117, .0000160, .0000161, .0000142, .0000114,$$
$$.854\,10^{-5}, .604\,10^{-5}, .402\,10^{-5}, .250\,10^{-5}, .143\,10^{-5}, .707\,10^{-6},$$
$$.257\,10^{-6}, .0311]$$

```
>  add(p.200[i,1]*(i-1) ,i = 1..dimP);
                 .4050783639
```

We make an animation of the distributions of the random variables X_N.

```
>  plots[display]([seq(
>  plots[pointplot]([seq([i-1,p.n[i,1]],i = 1..dimP)],
   connect =true, view = [0..dimP,0..1]), n = 0..200)],
   insequence = true):
```

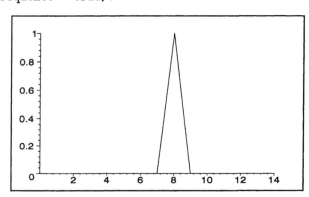

We urge the reader to look at the interesting animation on the computer screen. Alternatively, one can see the distributions for particular values of n.

```
>  plots[display]([seq(
   plots[pointplot]([seq([i-1,p.n[i,1]],i = 1..dimP)],
   connect = true, view = [0..dimP,0..1], tickmarks=[2,0],
   title = cat("after ",convert(n,string)," games")),
   n = [0,10,20,30,40,50,100,150,200] )], insequence = true):
>  plots[display](%):
```

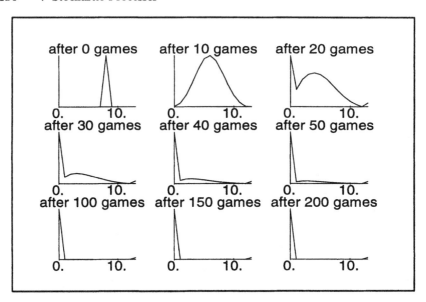

M 7.4 *Change the definition of the scaled random walk $S_N(t)$, (7.19), by extending its linearly over intervals. Make a plot.*

Modify the procedure Wpath used before.

```
>  Wpath := proc(steps,t)
>  local walk, i, N:
   N := nops(steps):
>  walk[0] := 0:
>  for i from 0 to N-1 do
   walk[i+1] := walk[i] + steps[i+1]*sqrt(t/N)
   od:
>  plots[pointplot]([seq([t*i/N,walk[i]],i = 0..N)],
   connect = true)
   end:
```

Make a plot. Note, that st[k] have to be defined as before.

```
>  plots[display]([seq(Wpath(st[k],4),k = [40,20,10,5,2,1])],
   insequence = true,axes = FRAMED, tickmarks = [2,3]):
>  plots[display](%);
```

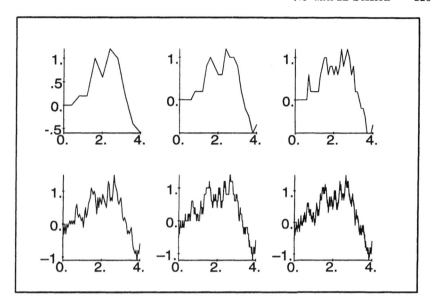

M 7.5 *Check that the transition density of the standard Wiener process satisfies the heat equation (7.23) and the equation (7.24).*

Define the transition density.
```
>  p := (s,x,t,y) ->
   1/sqrt(2*Pi*(t-s))*exp(-(y-x)^2/(t-s)/2);
```

$$p := (s, x, t, y) \rightarrow \frac{e^{(-1/2\frac{(y-x)^2}{t-s})}}{\sqrt{2\pi(t-s)}}$$

Verify the equations:
```
>  is(diff(p(s,x,t,y),t) - 1/2*diff(p(s,x,t,y),y,y) = 0);
```
$$true$$

```
>  is(diff(p(s,x,t,y),s) + 1/2*diff(p(s,x,t,y),x,x) = 0);
```
$$true$$

M 7.6 *Plot the above transition density as a function of (t,y), for $s = 0$ and $x = 0$.*

```
>  plot3d(p(0,0,t,y), t = 0..0.5, y = -1..1,axes = FRAMED,
   orientation = [200,40]#,style=wireframe); );
```

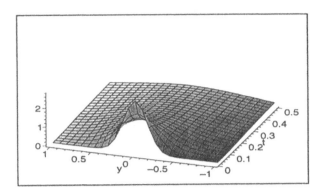

7.7 Exercises

E 7.1 Consider a random walk around the circle assuming that a particle can occupy only points corresponding to the angles $k\dfrac{2\pi}{n}$, when n is a given positive integer and $k = 0, \ldots, n - 1$. The particle moves clockwise with probability p, anti-clockwise with probability q and can stay at the same point with probability r, $p + q + r = 1$. Find the matrix $\mathbf{P}$ of the Markov chain.

E 7.2 Find the transition matrix and the initial distribution for the Markov chain described in Exercise E 1.11.

E 7.3 Let $\xi_0, \ldots, \xi_n$ be independent sequences of random variables taking integers values. Assume that ξ_0 has a distribution q_0 and all $\xi_1, \ldots, \xi_n$ have a common distribution q. Define

$$X_0 = \xi_0, \quad \text{and} \quad X_{n+1} = X_n + \xi_{n+1},$$

for $n = 0, 1, 2, \ldots$. Show that the random variables X_n form a Markov chain. Find its transition matrix and initial distribution.

E 7.4 Let $\mathbf{P}$ be the transition matrix for a Markov chain. Show that for all k the power matrix $\mathbf{P}^k$ satisfies the condition

$$\sum_{j \in E} \mathbf{P}^k(i, j) = 1, \text{ for each } i \in E.$$

E 7.5 Let X_n, $n = 0, 1, 2, \ldots$ be a Markov chain and let $k > 1$ be given. Show that the random variables $X_0, X_k, X_{2k}, \ldots$ also form a Markov chain.

E 7.6 Write MAPLE procedures to simulate Markov chains from Example 7.2.5 and Exercises 7.1 and 7.2.

E 7.7 Show that for ergodic Markov chains the distributions of the random variables X_n approach the stationary distribution π, i.e.

$$\lim_{n \to \infty} \mathbf{p}_n(i) = \pi(i), \text{ for each } i \in E,$$

for any initial distribution $\mathbf{p}_0$.

E 7.8 Find the stationary distributions, if they exist, for Markov chains from Example 7.2.5 and Exercises 7.1 and 7.2.

E 7.9 Give an example of an irreducible Markov chain with a finite collection of states E which is not ergodic.

E 7.10 Show that a standard Wiener process W has covariances $C(s, t) = \min\{s, t\}$ and hence is not wide-sense stationary.

E 7.11 Simulate a Brownian bridge $B_{0,x}^{T,y}(t)$. Assume for example: $x = -1$, $y = 1$, $T = 5$.

E 7.12 Prove that a Brownian bridge $B_{0,x}^{T,y}(t)$ for $0 \le t \le T$ is Gaussian with the means and covariances (7.17). Justify it experimentally too.

E 7.13 Show that the standard Wiener process is a diffusion process with drift $a(s, x) \equiv 0$ and diffusion coefficient $b(s, x) \equiv 1$, and that the Ornstein-Uhlenbeck process with $\gamma = 1$ is a diffusion process with drift $a(s, x) = -x$ and diffusion coefficient $b(s, x) \equiv \sqrt{2}$.

E 7.14 Prove property (7.37) for the telegraphic noise process.

E 7.15 Verify property (7.39) by explicitly determining the 4th moment of an $N(0, \sigma^2)$ distributed random variable.

8 Stochastic Calculus

8.1 Introduction

Stochastic calculus was introduced by the Japanese mathematician K. Ito during the 1940s. He was thus able to provide a mathematically meaningful formulation for stochastic differential equations, which until then had been treated only heuristically and inadequately.

Following Einstein's explanation of observed Brownian motion during the first decade of this 1900s, attempts were made by the physicists Langevin, Smoluchovsky and others to model the dynamics of such motion in terms of differential equations. Instead of a deterministic ordinary differential equation

$$\frac{dx}{dt} = a(t, x),$$

they obtained a noisy differential equation of the form

$$\frac{d}{dt} X_t = a(t, X_t) + b(t, X_t)\, \xi_t \tag{8.1}$$

with a deterministic or averaged drift term $a(t, X_t)$ perturbed by a noise term $b(t, X_t)\, \xi_t$, where the ξ_t are standard Gaussian random variables for each t and $b(t, x)$ is a intensity factor that may, in general, depend on both the t and x variables.

Observations indicated that the covariance $c(t) = \mathbb{E}(\xi_s \xi_{s+t})$ of the stochastic process ξ_t has a constant spectral density, hence with equally weighting of all time frequencies in any Fourier transform of $c(t)$, so $c(t)$ must be a constant multiple of the Dirac delta function $\delta(t)$. This means that the stochastic process ξ_t, which is known as *Gaussian white noise*, cannot be a conventional process. Indeed, taking $a \equiv 0$, $b \equiv 1$ in (8.1) above, implies that ξ_t should be the pathwise derivative of a mathematical Brownian motion or Wiener process W_t, i.e.

$$\xi_t(\omega) = \frac{d}{dt} W_t(\omega)$$

for each ω, which is problematic since the sample paths of a Wiener process W_t are nowhere differentiable!

Nevertheless, equation (8.1) might thus be rewritten symbolically in terms of differentials as

$$dX_t = a(t, X_t)\, dt + b(t, X_t)\, dW_t, \qquad (8.2)$$

which is to be interpreted in some sense as an integral equation

$$X_t(\omega) = X_{t_0}(\omega) + \int_{t_0}^{t} a(s, X_s(\omega))\, ds + \int_{t_0}^{t} b(s, X_s(\omega)) dW_s(\omega). \qquad (8.3)$$

However, the second integral in (8.3) cannot be a Riemann–Stieltjes integral for each sample path, since the continuous sample paths of a Wiener process are not even of bounded variation on any bounded time interval. Ito provided a means to overcome this difficulty with the definition of new type of integral, a stochastic integral that is now called an Ito stochastic integral.

8.2 Ito Stochastic Integrals

We want to define the stochastic integral

$$\int_{t_0}^{T} f(s, \omega) dW_s(\omega) \qquad (8.4)$$

for an appropriate random function $f : [0, T] \times \Omega \longrightarrow \mathbb{R}$, assuming for now that f is continuous in t for each $\omega \in \Omega$.

The Riemann–Stieltjes integral for a given sample path here would be the common limit of the sums

$$S_{(T,f)}(\omega) = \sum_{j=1}^{n} f\left(\tau_j^{(n)}, \omega\right) \left\{ W_{t_{j+1}^{(n)}}(\omega) - W_{t_j^{(n)}}(\omega) \right\} \qquad (8.5)$$

for all possible choices of evaluation points $\tau_j^{(n)} \in [t_j^{(n)}, t_{j+1}^{(n)}]$ and partitions

$$0 = t_1^{(n)} < t_2^{(n)} < \cdots < t_{n+1}^{(n)} = T$$

of $[0, T]$ as

$$\delta^{(n)} = \max_{1 \le j \le n} \left\{ t_{j+1}^{(n)} - t_j^{(n)} \right\} \longrightarrow 0 \quad \text{as} \quad n \longrightarrow \infty.$$

But, as mentioned above, such a limit does not exist due to the irregularity of sample paths of the Wiener process.

For mean-square convergence instead of pathwise convergence, the limit of the sums (8.5) may exist but take a different value depending on the particular choice of the evaluation points $\tau_j^{(n)} \in [t_j^{(n)}, t_{j+1}^{(n)}]$. For example, with $f(t, \omega) = W_t(\omega)$ and fixed evaluation point $\tau_j^{(n)} = (1 - \lambda)t_j^{(n)} + \lambda t_{j+1}^{(n)}$, where $\lambda \in [0, 1]$, for all $j = 0, 1, \ldots, n - 1$, the mean-square limit is equal to

$$\frac{1}{2}W_T^2 - \left(\frac{1}{2} - \lambda\right)T,$$

which we will denote by $(\lambda)\int_0^T W_t dW_t$. To see this, consider a constant step size $\delta = t_{j+1} - t_j = T/n$ (omitting the superscript n for convenience) and $\tau_j = (1-\lambda)t_j + \lambda t_{j+1} = (j+\lambda)\delta$. Then rearrange each term as

$$W_{\tau_j}\left\{W_{t_{j+1}} - W_{t_j}\right\} = -\frac{1}{2}\left(W_{t_{j+1}} - W_{\tau_j}\right)^2 + \frac{1}{2}\left(W_{\tau_j} - W_{t_j}\right)^2 + \frac{1}{2}\left(W_{\tau_j}^2 - W_{t_j}^2\right)$$

and take their sum to obtain

$$\sum_{j=1}^n W_{\tau_j}\left\{W_{t_{j+1}} - W_{t_j}\right\}$$

$$= -\frac{1}{2}\sum_{j=1}^n \left(W_{t_{j+1}} - W_{\tau_j}\right)^2 + \frac{1}{2}\sum_{j=1}^n \left(W_{\tau_j} - W_{t_j}\right)^2 + \frac{1}{2}\left(W_T^2 - W_0^2\right)$$

The first sum here converges in mean-square to $-\frac{1}{2}(1-\lambda)T$, the second to $\frac{1}{2}\lambda T$, and the third is equal to $\frac{1}{2}W_T^2$.

Ito proposed always to use the evaluation point $\tau_j = t_j$, that is with $\lambda = 0$. This has the advantage that the integrand value W_{τ_j} and the increment $W_{t_{j+1}} - W_{t_j}$ in each of the summation terms above are independent random variables, which together with the independence of nonoverlapping increments leads to

$$\mathbb{E}\left((1)\int_0^T W_t dW_t\right) = 0,$$

$$\mathbb{E}\left(\left|(1)\int_0^T W_t dW_t\right|^2\right) = \int_0^T \mathbb{E}\left(|W_t|^2\right)dt = \frac{1}{2}T^2.$$

The key observation here is that the integrand $f(t,\omega)$ (in this case $W_t(\omega)$) is independent of the future as represented by the increments of the Wiener process, that is f is a *nonanticipative stochastic process*; in addition, the second moment of the integrand, $\mathbb{E}\left(|f(t,\cdot)|^2\right)$, exists and is Riemann or Lebesgue integrable. Specifically, we define *nonanticipativeness* of the integrand f to mean that the random variable $f(t,\cdot)$ is $\mathcal{A}_t$-measurable for each $t \in [0,T]$, where $\{\mathcal{A}_t, t \geq 0\}$ is the increasing family of σ-algebras such that W_t is $\mathcal{A}_t$-measurable for each $t \geq 0$.

Thus we introduce a class of admissible integrand functions.

Definition 8.2.1 *The class $\mathcal{L}_T^2$ consists of functions $f : [0,T] \times \Omega \longrightarrow \mathbb{R}$ satisfying*

$$f \quad \text{is jointly } \mathcal{L} \times \mathcal{A}\text{-measurable;} \tag{8.6}$$

$$\int_0^T \mathbb{E}\left(f(t,\cdot)^2\right)\, dt < \infty; \tag{8.7}$$

$$\mathbb{E}\left(f(t,\cdot)^2\right) < \infty \quad \text{for each} \quad 0 \le t \le T; \tag{8.8}$$

and

$$f(t,\cdot) \text{ is } \mathcal{A}_t\text{-measurable for each} \quad 0 \le t \le T. \tag{8.9}$$

(We consider two functions in $\mathcal{L}_T^2$ to be identical if they are equal for all (t,ω) except possibly on a subset of $\mu_L \times \mathbb{P}$-measure zero).

Definition 8.2.2 *The Ito stochastic integral of $f \in \mathcal{L}_T^2$ is the unique mean-square limit*

$$\int_0^T f(t,\omega)dt = ms - \lim \sum_{j=1}^n f\left(t_j^{(n)},\omega\right)\left\{W_{t_{j+1}^{(n)}}(\omega) - W_{t_j^{(n)}}(\omega)\right\}. \tag{8.10}$$

Ito stochastic integrals enjoy the following properties.

Theorem 8.2.1 *The Ito stochastic integral (8.10) of $f \in \mathcal{L}_T^2$ satisfies*

$$I(f) \quad \text{is } \mathcal{A}_T\text{-measurable}, \tag{8.11}$$

$$\mathbb{E}(I(f)) = 0, \tag{8.12}$$

$$\mathbb{E}\left(I(f)^2\right) = \int_0^T \mathbb{E}\left(f(t,\cdot)^2\right)\, dt, \tag{8.13}$$

as well as

$$\mathbb{E}(I(f)I(g)) = \int_0^T \mathbb{E}(f(t,\cdot)g(t,\cdot))\, dt. \tag{8.14}$$

and the linearity property

$$I(\alpha f + \beta g) = \alpha I(f) + \beta I(g), \quad w.p.1 \tag{8.15}$$

for f, $g \in \mathcal{L}_T^2$ and any α, $\beta \in \mathbb{R}$.

Various applications require the Ito integral to be extended to a wider class of integrands than those in the space $\mathcal{L}_T^2$.

Definition 8.2.3 *A function $f : [0,T] \times \Omega \longrightarrow \mathbb{R}$ belongs to the class $\mathcal{L}_T^w$ if f is jointly $\mathcal{L} \times \mathcal{A}$-measurable, $f(t,\cdot)$ is $\mathcal{A}_t$-measurable for each $t \in [0,T]$ and*

$$\int_0^T f(s,\omega)^2\, ds < \infty, \quad w.p.1. \tag{8.16}$$

Hence $\mathcal{L}_T^2 \subset \mathcal{L}_T^w$. For $f \in \mathcal{L}_T^w$ we define $f_n \in \mathcal{L}_T^2$ by

$$f_n(t,\omega) = \begin{cases} f(t,\omega) & : \quad \text{if } \int_0^t f(s,\omega)^2\, ds \le n \\ \\ 0 & : \quad \text{otherwise} \end{cases}$$

The Ito stochastic integrals $I(f_n)$ of the f_n over $0 \le t \le T$ are thus well-defined for each $n \in \mathbb{N}$. It can then be shown that they converge in probability to a unique, w.p.1, random variable, which we denote by $I(f)$ and call the Ito stochastic integral of $f \in \mathcal{L}_T^w$ over the interval $[0, T]$. The Ito integrals of integrands $f \in \mathcal{L}_T^w$ satisfy analogous properties to those of integrands $f \in \mathcal{L}_T^2$, apart from those properties explicitly involving expectations, which may not exist now.

8.3 Stochastic Differential Equations

A stochastic differential equation (SDE)

$$dX_t = a(t, X_t)\, dt + b(t, X_t)\, dW_t \tag{8.17}$$

is interpreted as a stochastic integral equation

$$X_t = X_{t_0} + \int_{t_0}^t a(s, X_s)\, ds + \int_{t_0}^t b(s, X_s)\, dW_s, \tag{8.18}$$

where the first integral is a Lebesgue (or Riemann) integral for each sample path and the second integral is usually an Ito integral, in which case (8.18) is called an *Ito stochastic integral equation*, or Ito SDE for short.

A solution $\{X_t,\ t \in [t_0, T]\}$ of (8.18) must have properties which ensure that the stochastic integral is meaningful, i.e. so its integrand belongs to the spaces $\mathcal{L}_T^2$ or $\mathcal{L}_T^w$ defined in the previous section. This follows if the coefficient function b is sufficiently regular and if the process X_t is *nonanticipative* with respect to the given Wiener process $\{W_t, t \ge 0\}$, i.e. $\mathcal{A}_t$–measurable for each $t \ge 0$. Then we call the process $X = \{X_t,\ t_0 \le t \le T\}$ a *solution* of (8.18) on $[t_0, T]$. Moreover, we say that solutions $\tilde{X}_t$ and X_t of (8.18) are *pathwise unique* if they have, w.p.1, the same sample paths, i.e. if

$$P\left(\sup_{0 \le t \le T} \left| \tilde{X}_t - X_t \right| = 0 \right) = 1.$$

The assumptions of an existence and uniqueness theorem are usually sufficient, but not necessary, conditions to ensure the conclusion of the theorem. Those stated below for scalar SDE are quite strong, but can be weakened in several ways. In what follows the initial instant $t_0 \in [0, T]$ is arbitrary, but fixed, and the coefficients functions $a,\ b : [t_0, T] \times \mathbb{R} \longrightarrow \mathbb{R}$ are given, as is the Wiener process $\{W_t, t \geq 0\}$ and its associated family of σ-algebras $\{\mathcal{A}_t, t \geq 0\}$. Most of the assumptions concern the coefficients.

A1 (Measurability): $a = a(t,x)$ and $b = b(t,x)$ are jointly ($\mathcal{L}^2$-) measurable in $(t,x) \in [t_0, T] \times \mathbb{R}$;

A2 (Lipschitz condition):   There exists a constant $K > 0$ such that

$$|a(t,x) - a(t,y)| \leq K\,|x - y|$$

and

$$|b(t,x) - b(t,y)| \leq K\,|x - y|$$

for all $t \in [t_0, T]$ and $x, y \in \mathbb{R}$;

A3 (Linear growth bound):   There exists a constant $K > 0$ such that

$$|a(t,x)|^2 \leq K^2 (1 + |x|^2)$$

and

$$|b(t,x)|^2 \leq K^2 (1 + |x|^2)$$

for all $t \in [t_0, T]$ and $x, y \in \mathbb{R}$.

A4 (Initial value): X_{t_0} is $\mathcal{A}_{t_0}$-measurable with $E(|X_{t_0}|^2) < \infty$.

The Lipschitz condition A2 provides the key estimates in both the proofs of uniqueness and of existence by the method of successive approximations.

Theorem 8.3.1 *Under assumptions A1–A4 the stochastic differential equation (8.18) has a pathwise unique solution X_t on $[t_0, T]$ with*

$$\sup_{t_0 \leq t \leq T} \mathbb{E}\left(|X_t|^2\right) < \infty.$$

The proof is by the method of successive approximations, first defining $X_t^{(0)} \equiv X_{t_0}$ and then

$$X_t^{(n+1)} = X_{t_0} + \int_{t_0}^{t} a\left(s, X_s^{(n)}\right)\,ds + \int_{t_0}^{t} b\left(s, X_s^{(n)}\right)\,dW_s \qquad (8.19)$$

for $n = 0, 1, 2, \ldots$.

The main difference from the deterministic case is that we have to estimate stochastic integrals. For this we consider the squares of all expressions and then take the expectation, as this converts the Ito integral into a deterministic integral of the expectation of the squared integrand, which we can easily estimate. For example, from assumption A3 and the definition of $X_t^{(0)}$ one can show that

$$\mathbb{E}\left(\left|\int_{t_0}^t b\left(s, X_s^{(n)}\right) dW_s\right|^2\right) = \int_{t_0}^t \mathbb{E}\left(\left|b\left(s, X_s^{(n)}\right)\right|^2\right) ds$$

$$\leq K^2 \int_{t_0}^t \left(1 + \mathbb{E}\left(\left|X_s^{(n)}\right|^2\right)\right) ds$$

for $n = 0, 1, 2, \ldots$. A similar procedure is used to exploit the Lipschitz property of the b coefficient.

From their construction, the successive approximations $X_t^{(n)}$ above are obviously $\mathcal{A}_t$-measurable and have, almost surely, continuous sample paths. These properties are inherited by a limiting solution X_t of the SDE. It is not difficult to show the mean-square convergence of the successive approximations and the mean-square uniqueness of the limiting solution. The Borel-Cantelli Lemma is then applied to establish the stronger pathwise uniqueness of the solutions. The details of this part of the proof are technically complicated. See Kloeden and Platen [19], for example.

An important property of the solutions of an Ito SDE (8.17) is that they are Markov processes and, in fact, often diffusion processes. When the coefficients a and b of the SDE are sufficiently smooth, the transition probabilities of this Markov process have a density $p = p(s, x; t, y)$ satisfying the *Fokker-Planck equation*

$$\frac{\partial p}{\partial t} + \frac{\partial}{\partial y}(ap) - \frac{1}{2}\frac{\partial^2}{\partial y^2}(\sigma p) = 0 \qquad (8.20)$$

with $\sigma = b^2$.

Example 8.3.1 The Fokker-Planck equation corresponding to the Ito SDE

$$dX_t = -X_t dt + \sqrt{2} dW_t$$

is

$$\frac{\partial p}{\partial t} + \frac{\partial}{\partial y}(-yp) - \frac{\partial^2 p}{\partial y^2} = 0$$

and has the solution

$$p(s, x; t, y) = \frac{1}{\sqrt{2\pi\left(1 - e^{-2(t-s)}\right)}} \exp\left(-\frac{\left(y - xe^{-(t-s)}\right)^2}{2\left(1 - e^{-2(t-s)}\right)}\right)$$

MAPLE ASSISTANCE 79

We define a procedure FPE that constructs the appropriate Fokker-Planck equation corresponding to a given scalar Ito SDE.

```
>   FPE := proc(a,b)
>   Diff(p,t)+Diff(a*p,y)-Diff(1/2*b^2*p,y,y)=0;
>   subs(x=y,%):
>   end:
```

This procedure can be applied to any scalar Ito SDE. Here we apply it to the SDE in the above example.

```
>  FPE(-x,sqrt(2));
```

$$(\frac{\partial}{\partial t} p) + (\frac{\partial}{\partial y} (-y \, p)) - (\frac{\partial^2}{\partial y^2} p) = 0$$

Note that the input for the above procedure must be in the variable x.

8.4 Stochastic Chain Rule: the Ito Formula

Although the Ito stochastic integral has some very convenient properties as listed in Theorem 8.2.1, it also has some peculiar features compared to deterministic calculus, which the value of the Ito integral $\int_0^T W_t dW_t$ and the solution of the Ito SDE $dX_t = X_t dW_t$ given in the previous sections already indicate. These peculiarities are a direct consequence of the fact that the solutions of Ito SDE do not transform according to the usual chain rule of classical calculus. Instead, the appropriate stochastic chain rule, known as the *Ito formula*, contains an additional term, which, roughly speaking, is due to the fact that the stochastic differential $(dW_t)^2$ is equal to dt in the mean-square sense, i.e. $\mathbb{E}((dW_t)^2) = dt$, so the second order term in dW_t should really appear as a first order term in dt.

Let $Y_t = U(t, X_t)$, where $U : [t_0, T] \times \mathbb{R} \longrightarrow \mathbb{R}$ has continuous second order partial derivatives, and let X_t be a solution of the Ito SDE

$$dX_t = a(t, X_t) \, dt + b(t, X_t) \, dW_t.$$

Then differential of Y_t satisfies the equation

$$dY_t = L^0 U(t, X_t) \, dt + L^1 U(t, X_t) \, dW_t, \tag{8.21}$$

where the partial differential operators L^0 and L^1 are defined as

$$L^0 U = \frac{\partial U}{\partial t} + a \frac{\partial U}{\partial x} + \frac{1}{2} b^2 \frac{\partial^2 U}{\partial x^2} \tag{8.22}$$

and

$$L^1 = b \frac{\partial U}{\partial x}, \tag{8.23}$$

where all functions are evaluated at (t, x). This is known as the *Ito formula* and interpreted as the Ito stochastic integral equation

$$Y_t = Y_{t_0} + \int_{t_0}^t L^0 U(s, X_s) \, ds + \int_{t_0}^t L^1 U(s, X_s) \, dW_s$$

for $t \in [t_0, T]$.

The final term in the definition (8.22) of the L^0 operator is not present in the deterministic chain rule. To see how it arises, consider the first few terms in the deterministic Taylor expansion for U,

$$\Delta Y_t = U\left(t + \Delta t, X_t + \Delta X_t\right) - U\left(t, X_t\right)$$

$$= \left\{ \frac{\partial U}{\partial t} \Delta t + \frac{\partial U}{\partial x} \Delta x \right\}$$

$$+ \frac{1}{2} \left\{ \frac{\partial^2 U}{\partial t^2} (\Delta t)^2 + 2 \frac{\partial^2 U}{\partial t \partial x} \Delta t \, \Delta x + \frac{\partial^2 U}{\partial x^2} (\Delta X_t)^2 \right\} + \cdots$$

where the partial derivatives are evaluated at (t, X_t) and the dots indicate higher order terms. Now $\Delta X_t = a\Delta t + b\Delta W_t + \cdots$, so

$$(\Delta X_t)^2 = a^2 (\Delta t)^2 + 2ab\Delta t \Delta W_t + b^2 (\Delta W_t)^2 + \cdots,$$

where the coefficient functions a and b are evaluated at (t, X_t). Thus we obtain

$$\Delta Y_t = \left\{ \frac{\partial U}{\partial t} + a \frac{\partial U}{\partial x} \right\} \Delta t + \frac{1}{2} b^2 \frac{\partial^2 U}{\partial x^2} (\Delta W_t)^2 + b \frac{\partial U}{\partial x} \Delta W_t + \cdots$$

$$= \left\{ \frac{\partial U}{\partial t} + a \frac{\partial U}{\partial x} + \frac{1}{2} b^2 \frac{\partial^2 U}{\partial x^2} \right\} \Delta t + \frac{\partial U}{\partial x} \Delta W_t + \cdots$$

where the equality in the last line is interpreted in the mean-square sense. Replacing the increments by differentials and discarding the higher order terms gives the Ito formula (8.21).

Example 8.4.1 For $X_t = W_t$ and $Y_t = X_t^2$, so $a \equiv 0$, $b \equiv 1$ and $U(t, x) = x^2$, the Ito formula gives

$$dY_t = d\left(X_t^2\right) = 1\, dt + 2X_t\, dW_t.$$

Hence

$$W_t\, dW_t = \frac{1}{2} d\left(W_t^2\right) - \frac{1}{2} dt$$

and so in integral form

$$\int_0^t W_s\, dW_s = \frac{1}{2} \int_0^t d\left(W_s^2\right) - \frac{1}{2} \int_0^t 1\, ds$$

$$= \frac{1}{2} W_t^2 - \frac{1}{2} t,$$

since $W_0 = 0$, w.p.1.

MAPLE ASSISTANCE 80

Below we present the procedure *Ito*, which applies the Ito formula (8.21). First, however, two procedures L0 and L1 are given, which apply the operators L^0 and L^1, respectively. These will be required for procedure Ito.

```
>  L0 := proc(X,a,b)
>  local Lzero,U;
>  Lzero := diff(U(x,t),t)+a*diff(U(x,t),x)+
   1/2*b^2*diff(U(x,t),x,x);
>  eval(subs(U(x,t)=X,Lzero));
>  end:
>  L1 := proc(X,b)
>  local Lone,U;
>  Lone := b*diff(U(x,t),x);
>  eval(subs(U(x,t)=X,Lone));
>  end:
```

Now we define Ito.

```
>  Ito := proc(Y,a,b)
>  local soln;
>  soln := dY=L0(Y,a,b)*dt+L1(Y,b)*dW;
>  subs(x=X,soln);
>  end:
```

Next we apply this procedure to the above example.

```
>  Ito(x^2,0,1);
```

$$dY = dt + 2X\, dW$$

Note that this procedure may be used for any given Y_t. There are three input parameters; Y_t, a and b, respectively. Again, parameters should be given in terms of x.

8.5 Stochastic Taylor Expansions

Stochastic Taylor expansions generalize the deterministic Taylor formula. They are based on an iterated application of the Ito formula and provide various kinds of higher order approximations of functions of the solutions processes of Ito SDEs, which are the key to developing higher order numerical schemes for SDEs.

We indicate the derivation of stochastic Taylor expansions here for the solution X_t of the scalar Ito stochastic differential equation in integral form

$$X_t = X_{t_0} + \int_{t_0}^t a(s, X_s)\, ds + \int_{t_0}^t b(s, X_s)\, dW_s \qquad (8.24)$$

for $t \in [t_0, T]$, where the second integral is an Ito stochastic integral and the coefficients a and b are sufficiently smooth real-valued functions satisfying a linear growth bound. Then, for any twice continuously differentiable function $f : \mathbb{R} \times \mathbb{R} \longrightarrow \mathbb{R}$ the Ito formula (8.21) gives

$$f(t, X_t) = f(t_0, X_{t_0}) + \int_{t_0}^t L^0 f(s, X_s) \, ds + \int_{t_0}^t L^1 f(s, X_s) \, dW_s, \quad (8.25)$$

for $t \in [t_0, T]$. The operators L^0 and L^1 here were defined in (8.22) and (8.23), respectively, in the previous section.

Obviously, for $f(t, x) \equiv x$ we have $L^0 f = a$ and $L^1 f = b$, in which case (8.25) reduces to the original Ito equation for X_t, that is to

$$X_t = X_{t_0} + \int_{t_0}^t a(s, X_s) \, ds + \int_{t_0}^t b(s, X_s) \, dW_s. \quad (8.26)$$

If we apply the Ito formula (8.21) to each of the functions $f = a$ and $f = b$ in the integrands of (8.26) we obtain

$$X_t = X_{t_0} + \int_{t_0}^t \left\{ a(t_0, X_{t_0}) + \int_{t_0}^s L^0 a(r, X_r) \, dr \, ds + \int_{t_0}^s L^1 a(r, X_r) \, dW_r \right\} ds$$
$$+ \int_{t_0}^t \left\{ b(t_0, X_{t_0}) + \int_{t_0}^s L^0 b(r, X_r) \, dr \, ds + \int_{t_0}^s L^1 b(r, X_r) \, dW_r \right\} dW_s,$$

which gives the expansion

$$X_t = X_{t_0} + a(t_0, X_{t_0}) \int_{t_0}^t ds + b(t_0, X_{t_0}) \int_{t_0}^t dW_s + R \quad (8.27)$$

with the remainder

$$R = \int_{t_0}^t \int_{t_0}^s L^0 a(r, X_r) \, dr \, ds + \int_{t_0}^t \int_{t_0}^s L^1 a(r, X_r) \, dW_r \, ds$$
$$+ \int_{t_0}^t \int_{t_0}^s L^0 b(r, X_r) \, dr \, dW_s + \int_{t_0}^t \int_{t_0}^s L^1 b(r, X_r) \, dW_r \, dW_s.$$

This is the simplest nontrivial stochastic Taylor expansion. We can continue it, for instance, by applying the Ito formula (8.21) to $f = L^1 b$, in which case we get

$$X_t = X_{t_0} + a(t_0, X_{t_0}) \int_{t_0}^t ds + b(t_0, X_{t_0}) \int_{t_0}^t dW_s$$
$$+ L^1 b(t_0, X_{t_0}) \int_{t_0}^t \int_{t_0}^s dW_r \, dW_s + \bar{R} \quad (8.28)$$

with remainder

$$\bar{R} = \int_{t_0}^{t} \int_{t_0}^{s} L^0 a(r, X_r)\, dr\, ds + \int_{t_0}^{t} \int_{t_0}^{s} L^1 a(r, X_r)\, dW_r\, ds$$

$$+ \int_{t_0}^{t} \int_{t_0}^{s} L^0 b(r, X_r)\, dr\, dW_s + \int_{t_0}^{t} \int_{t_0}^{s} \int_{t_0}^{r} L^0 L^1 b(u, X_u)\, du\, dW_r\, dW_s$$

$$+ \int_{t_0}^{t} \int_{t_0}^{s} \int_{t_0}^{r} L^1 L^1 b(u, X_u)\, dW_u\, dW_r\, dW_s.$$

The main properties are already apparent in the preceding example. In particular, we have an expansion with the multiple Ito integrals

$$\int_{t_0}^{t} ds, \qquad \int_{t_0}^{t} dW_s, \qquad \int_{t_0}^{t} \int_{t_0}^{s} dW_r\, dW_s$$

and a remainder term involving the next multiple Ito integrals, but now with nonconstant integrands. The Ito-Taylor expansion can, in this sense, be interpreted as a generalization of both the Ito formula and the deterministic Taylor formula. In fact, there are many possible Ito-Taylor expansions since we have a choice of which of the integrals in the remainder term we will expand at the next step.

Example 8.5.1 Let X_t be a solution of bilinear SDE

$$dX_t = aX_t\, dt + bX_t\, dW_t$$

and let $U(t, x) = x^2$. Then

$$L^0 U = (2a + b^2)x^2, \qquad L^1 U = 2bx^2, \qquad L^1 L^1 U = 4b^2 x^2,$$

so the Taylor expansion (8.28) in this case is

$$X_t^2 = X_{t_0}^2 + (2a + b^2) X_{t_0}^2 \int_{t_0}^{t} ds + 2b X_{t_0}^2 \int_{t_0}^{t} dW_s$$

$$+ 4b^2 X_{t_0}^2 \int_{t_0}^{t} \int_{t_0}^{s} dW_r\, dW_s + \bar{R}$$

with remainder $\bar{R}$.

MAPLE ASSISTANCE 81

Here we make use of procedures L0 and L1 defined previously. Applying these procedures to this example we can quickly obtain the above results. Defining such procedures in MAPLE is useful in instances such as these, especially when many applications of the operators L^0 and L^1 are required.

```
> L0(x^2,a*x,b*x);
```

$$2\, a\, x^2 + b^2\, x^2$$

```
>  factor(%);
```
$$x^2 \left(2\,a + b^2 \right)$$

```
>  L1(x^2,b*x);
```
$$2\,b\,x^2$$

```
>  L1(L1(x^2,b*x),b*x);
```
$$4\,b^2\,x^2$$

Note the first parameter required for both L0 and L1 is U. For L0 the remaining parameters required are a and b, while L1 requires only b. Again, input must be in terms of x.

8.6 Stratonovich Stochastic Calculus

There is another stochastic integral that was introduced by the Russian physicist R.L. Stratonovich in the 1960s. It differs from the definition of the Ito stochastic integral (8.10) in its use of the midpoint $\tau_j^{(n)} = \frac{1}{2}(t_j^{(n)} + t_{j+1}^{(n)})$ of each partition subinterval $[t_j^{(n)}, t_{j+1}^{(n)}]$ as the evaluation point of the integrand instead of the lower end point $t_j^{(n)}$.

The *Stratonovich integral* of an integrand $f \in \mathcal{L}_T^2$ is defined as the mean–square limit

$$\int_0^T f(t,\omega) \circ dW_t = ms - \lim \sum_{j=1}^n f\left(\tau_j^{(n)}, \omega\right) \left\{ W_{t_{j+1}^{(n)}}(\omega) - W_{t_j^{(n)}}, (\omega) \right\} \quad (8.29)$$

where $\tau_j^{(n)}$ is the above midpoint. The symbol $\circ$ is used here to distinguished the Stratonovich stochastic integral from the Ito stochastic integral. In particular,

$$\int_0^T W_t \circ dW_t = \frac{1}{2} W_T^2,$$

just like a deterministic integral, in contrast to the corresponding Ito integral

$$\int_0^T W_t dW_t = \frac{1}{2} W_T^2 - \frac{1}{2} T.$$

Usually only the Ito and Stratonovich integrals are widely used. The Stratonovich integral obeys the transformation rules of clasical calculus, and this is a major reason for its use, but it does not have some of the convenient properties of the mean and variance of an Ito integral.

If the Stratonovich stochastic integral is used in the SDE instead of the Ito stochastic integral as in the Ito SDE (8.18) above, we obtain a *Stratonovich stochastic integral equation*, which we write symbolically as

$$dX_t = a(t, X_t)\, dt + b(t, X_t) \circ dW_t \tag{8.30}$$

and interpret as the Stratonovich stochastic integral equation

$$X_t = X_{t_0} + \int_{t_0}^t a(s, X_s)\, ds + \int_{t_0}^t b(s, X_s) \circ dW_s. \tag{8.31}$$

Ito and Stratonovich SDEs with the same coefficients usually do not have the same solution. For example (see Exercise **E** 8.4) the solution of the Stratonovich SDE $dX_t = X_t \circ dW_t$ with the initial value $X_0 = 1$ is $X_t = \exp(W_t)$, whereas the solution of the Ito SDE $dX_t = X_t\, dW_t$ for the same initial value is $X_t = \exp\left(W_t - \frac{1}{2}t\right)$.

However, the solution X_t of an Ito SDE

$$X_t = X_{t_0} + \int_{t_0}^t a(s, X_s)\, ds + \int_{t_0}^t b(s, X_s)\, dW_s \tag{8.32}$$

satisfies the *modified* Stratonovich SDE

$$X_t = X_{t_0} + \int_{t_0}^t \underline{a}(s, X_s)\, ds + \int_{t_0}^t b(s, X_s) \circ dW_s \tag{8.33}$$

with the modified drift coefficient

$$\underline{a}(t, x) = a(t, x) - \frac{1}{2} b(t, x) \frac{\partial b}{\partial x^k}(t, x) \tag{8.34}$$

and the same diffusion coefficient $b(t, x)$. This enables us to switch from one type of SDE to the other to use their respective properties to our advantage. It is, however, a major modelling problem, essentially unresolved, to decide whether the Ito or Stratonovich version of an SDE should be used to start with.

Example 8.6.1 The Ito SDE

$$dX_t = aX_t\, dt + bX_t\, dW_t$$

has the same solution as the Stratonovich SDE

$$dX_t = \left(a - \frac{1}{2} b^2\right) X_t\, dt + bX_t \circ dW_t,$$

whereas the Stratonovich SDE

$$dX_t = aX_t\, dt + bX_t \circ dW_t$$

has the same solution as the S Ito SDE

$$dX_t = \left(a + \frac{1}{2} b^2\right) X_t\, dt + bX_t\, dW_t.$$

MAPLE ASSISTANCE 82

Here we give procedures ItoS and StoI, which convert an Ito SDE to a Stratonovich SDE and a Stratonovich SDE to an Ito SDE, respectively.

```
>   ItoS:=proc(a,b)
>   local moda;
>   moda:=a-1/2*b*diff(b,x);
>   dX=moda*dt+b*SdW;
>   subs(x=X,%);
>   end:
>   StoI:=proc(moda,b)
>   local a;
>   a:=moda+1/2*b*diff(b,x);
>   dX=a*dt+b*dW;
>   subs(x=X,%);
>   end:
```

Next, we apply these procedures to the above example. Two input parameters are required for these procedures; drift a and diffusion b.

```
>   ItoS(a*x,b*x);
```

$$dX = \left(a\,X - \frac{1}{2}\,b^2\,X\right) dt + b\,X\,SdW$$

```
>   StoI(a*x,b*x);
```

$$dX = \left(a\,X + \frac{1}{2}\,b^2\,X\right) dt + b\,X\,dW$$

The output for procedure ItoS includes the term SdW to denote, in MAPLE, that we have a Stratonovich SDE. Note again that input must be in terms of x.

$$\cdots$$

When X_t is a solution of a Stratonovich SDE

$$dX_t = a(t, X_t)\,dt + b(t, X_t) \circ dW_t,$$

the transformed process $Y_t = U(t, X_t)$ satisfies the equation

$$dY_t = \underline{L}^0 U(t, X_t)\,dt + \underline{L}^1 U(t, X_t) \circ dW_t, \qquad (8.35)$$

which is interpreted as the Stratonovich stochastic integral equation

$$Y_t = Y_{t_0} + \int_{t_0}^t \underline{L}^0 U(s, X_s)\,ds + \int_{t_0}^t \underline{L}^1 U(s, X_s) \circ dW_s,$$

for $t \in [t_0, T]$, where the partial differential operators $\underline{L}^0$ and $\underline{L}^1$ are now defined as

$$\underline{L}^0 U = \frac{\partial U}{\partial t} + a \frac{\partial U}{\partial x} \tag{8.36}$$

and

$$\underline{L}^1 = b \frac{\partial U}{\partial x}, \tag{8.37}$$

and all functions are evaluated at (t, x). Note in particular, that the $\underline{L}^0$ operator does not contain the final term of the L^0 operator in the Ito case. In fact, the Stratonovich transformation rule (8.35) is exactly what one would obtain using deterministic calculus with a continuously differentiable function $w(t)$ instead of a sample path of a Wiener process. We will see that this means deterministic substitution tricks can be used to solve Stratonovich SDEs.

Example 8.6.2 For $X_t = W_t$, which satisfies the Stratonovich SDE $dX_t = 1 \circ dW_t$ with $a \equiv 0$ and $b \equiv 1$, and for $Y_t = X_t^2$, so $U(t, x) = x^2$, the Stratonovich transformation rule (8.35) gives

$$dY_t = d\left(X_t^2\right) = 0 \, dt + 2X_t \circ dW_t = 2W_t \circ dW_t,$$

hence

$$\int_0^t W_s \circ dW_s = \frac{1}{2} W_t^2$$

since $W_0 = 0$, w.p.1.

8.7 MAPLE Session

M 8.1 *Show that* $d(\cos W_t) = -\frac{1}{2} \cos W_t dt - \sin W_t dW_t$ *and* $d(\sin W_t) = -\frac{1}{2} \sin W_t dt + \cos W_t dW_t$.

Here we apply the previously defined procedure Ito to $Y_t = U(X_t)$ with $U(x) = \cos x$ and $X_t = W_t$, that is with $a = 0$ and $b = 1$.

```
>  Ito(cos(x),0,1);
```
$$dY = -\frac{1}{2} \cos(X) \, dt - \sin(X) \, dW$$

Substitution of $X_t = W_t$ then yields the desired result.

Repeating this process for $Y_t = U(X_t)$ with $U(x) = \sin x$ and $X_t = W_t$ we obtain the second result.

```
>  Ito(sin(x),0,1);
```
$$dY = -\frac{1}{2} \sin(X) \, dt + \cos(X) \, dW$$

Again, substitution of $X_t = W_t$ yields the desired result.

M 8.2 *Determine and solve the Stratonovich SDE corresponding to the Ito SDE $dX_t = 1dt + 2\sqrt{X_t}dW_t$.*

First, we use the previously defined procedure ItoS to convert the Ito SDE to the corresponding Stratonovich SDE.

```
>  ItoS(1,2*sqrt(x));
```
$$dX = 2\sqrt{X}\,SdW$$

Then, we rewrite this SDE so that MAPLE recognises it as a differential equation obeying the rules of deterministic calculus.

```
>  diff(X(W),W)=2*sqrt(X(W));
```
$$\frac{\partial}{\partial W}X(W) = 2\sqrt{X(W)}$$

Next, we ask MAPLE to solve this differential equation.

```
>  factor(dsolve(%,X(W)));
```
$$X(W) = (W + \_C1)^2$$

M 8.3 *Determine and solve the Stratonovich SDE corresponding to the Ito SDE $dX_t = \frac{1}{3}X_t^{\frac{1}{3}}dt + X_t^{\frac{2}{3}}dW_t$.*

Repeating the steps for the previous exercise, we first use procedure ItoS to convert the Ito SDE to the corresponding Stratonovich SDE.

```
>  ItoS(1/3*x^(1/3),x^(2/3));
```
$$dX = X^{(2/3)}\,SdW$$

Next, rewrite the SDE.

```
>  diff(X(W),W)=X(W)^(2/3);
```
$$\frac{\partial}{\partial W}X(W) = X(W)^{(2/3)}$$

And ask MAPLE to solve.

```
>  factor(dsolve(%,X(W)));
```
$$X(W) = \frac{1}{27}(W + 3\_C1)^3$$

This last SDE has nonunique solutions for $X_0 = 0$; for example, $X_t \equiv 0$ is a solution in addition to the solution $X_t = \frac{1}{27} W_t^3$ obtained from the above procedure.

M 8.4 $dX_t = -(a \sin X_t + 2b \sin 2X_t)dt + \sigma dW_t$ *is an SDE that is used in experimental psychology to describe the coordinated movement of a subject's left and right index fingers. For this SDE, calculate the result of the operator combination $L^1 L^0$ applied to the function $f(x) = x$.*

We will apply the previously defined procedures L0 and L1 to x, using the drift and diffusion coefficients from the above SDE.

```
>  L1(L0(x,-(a*sin(x)+2*b*sin(2*x)),sigma),sigma);
```
$$\sigma\left(-a\cos(x) - 4\,b\cos(2\,x)\right)$$

We will see in the final chapter that such operator combinations are common in stochastic numerics.

8.8 Exercises

E 8.1 Use the Ito formula to show that

$$d\left(X_t^{2n}\right) = n(2n-1)\, f_t^2\, X_t^{2n-2}\, dt + 2n\, f_t\, X_t^{2n-1}\, dW_t$$

for $n \geq 1$, where $dX_t = f_t\, dW_t$, where f is a nonanticipative function. Hence determine $d\left(W_t^{2n}\right)$ for $n \geq 1$.

E 8.2 Let $dX_t = f_t\, dW_t$, where f is a nonanticipative function, and $Y_t = U(t, X_t)$ with $U(t, x) = e^x$. Use the Ito formula to show that

$$dY_t = \frac{1}{2}\, f_t^2\, Y_t\, dt + f_t\, Y_t\, dW_t,$$

whereas with $U(t, x) = \exp\left(x - \frac{1}{2}\int_0^t f_u^2\, du\right)$ it gives

$$dY_t = f_t\, Y_t\, dW_t.$$

E 8.3 Use ordinary calculus to solve the Stratonovich SDE

$$dY_t = \exp\left(-Y_t\right) \circ dW_t$$

for $Y_0 = 0$. Hence determine the Ito SDE that is satisfied by this solution.

E 8.4 Show that $Y_t = \exp(W_t)$ is a solution of the Stratonovich SDE

$$dY_t = Y_t \circ dW_t,$$

whereas $Y_t = \exp\left(W_t - \frac{1}{2}t\right)$ is the solution of the Ito SDE

$$dY_t = Y_t \, dW_t$$

for the same initial value $Y_0 = 1$.

E 8.5 Show that for an $r + 1$ times continuously differentiable function $U :$ $\mathbb{R} \times \mathbb{R} \longrightarrow \mathbb{R}$, the stochastic Taylor expansion for $U(t, x(t))$ for a solution $x(t)$ of the ordinary differential equation

$$\frac{dx}{dt} = a(t, x)$$

reduces to the deterministic Taylor expansion (in integral form)

$$U(t, x(t)) = U(t_0, x(t_0)) + \sum_{l=1}^{r} \frac{1}{l!} D^l U(t_0, x(t_0)) (t - t_0)^l$$
$$+ \underbrace{\int_{t_0}^{t} \cdots \int_{t_0}^{s_2} D^{r+1} U(s_1, x(s_1)) \, ds_1 \ldots ds_{r+1}}_{r+1 \text{ times}}$$

for $t \in [t_0, T]$ and any $r = 1, 2, 3, \ldots$, where D is the total derivative defined by

$$D = \frac{\partial}{\partial t} + a \frac{\partial}{\partial x}.$$

(Hint: $L^0 \equiv D$ and $L^1 \equiv 0$ when $b \equiv 0$ in an Ito SDE):

E 8.6 Show that for a general $r + 1$ times continuously differentiable function $U : \mathbb{R} \longrightarrow \mathbb{R}$ the stochastic Taylor expansion reduces to the classical *Taylor formula in integral form*:

$$U(t) = U(t_0) + \sum_{l=1}^{r} \frac{1}{l!} \frac{d^l U}{dt^l}(t_0) (t - t_0)^l + \underbrace{\int_{t_0}^{t} \cdots \int_{t_0}^{s_2} \frac{d^{r+1} U}{dt^{r+1}}(s_1) \, ds_1 \ldots ds_{r+1}}_{r+1 \text{ times}}$$

for $t \in [t_0, T]$ and any $r = 1, 2, 3, \ldots$.

9 Stochastic Differential Equations

Techniques for solving linear and certain classes of nonlinear stochastic differential equations are presented along with MAPLE routines. Scalar SDEs will be considered first and then, at the end of the chapter, vector SDEs.

9.1 Solving Scalar Stratonovich SDEs

We have seen that a scalar Ito SDE

$$dX_t = a(t, X_t)\, dt + b(t, X_t)\, dW_t \tag{9.1}$$

has the same solution as the scalar Stratonovich SDE

$$dX_t = \underline{a}(t, X_t)\, dt + b(t, X_t) \circ dW_t \tag{9.2}$$

with the modified drift $\underline{a}$ defined by

$$\underline{a}(t, x) = a(t, x) - \frac{1}{2} b(t, x) \frac{\partial b}{\partial x}(t, x).$$

Moreover we have seen that the transformations of solutions of a Stratonovich SDE satisfy the same chain rule as in deterministic calculus. This means we can use the same integration substitution tricks solve from deterministic calculus to solve Stratonovich SDEs.

Example 9.1.1 Consider the bilinear Stratonovich SDE

$$dX_t = aX_t\, dt + bX_t \circ dW_t \tag{9.3}$$

Dividing both sides by X_t gives

$$\frac{1}{X_t}\, dX_t = a\, dt + b \circ dW_t = a\, dt + b\, dW_t,$$

which we integrate from t_0 to t to obtain

$$\int_{t_0}^{t} \frac{1}{X_s}\, dX_s = \int_{t_0}^{t} a\, ds + \int_{t_0}^{t} b\, dW_s,$$

which gives

$$\ln\left(\frac{X_t}{X_{t_0}}\right) = a(t - t_0) + b(W_t - W_{t_0}),$$

and hence

$$X_t = X_{t_0} \exp\left(a(t - t_0) + b(W_t - W_{t_0})\right). \tag{9.4}$$

Here we have used the substitution $u = \ln x$ to obtain

$$\int_{t_0}^{t} \frac{1}{X_s} dX_s = \int_{\ln X_{t_0}}^{\ln X_t} 1 \, du = \ln X_t - \ln X_{t_0} = \ln\left(\frac{X_t}{X_{t_0}}\right).$$

Note that the Ito SDE

$$dX_t = \left(a + \frac{1}{2} b^2\right) X_t \, dt + b X_t \, dW_t$$

has equation (9.3) as its modified Stratonovich SDE, so its solution is also given by (9.4).

Certain types of nonlinear scalar SDE can also be integrated explicitly in the same way. We consider two classes here, starting first with the Ito version of the SDE.

Reducible Case 1

The Ito SDE

$$dX_t = \frac{1}{2} b(X_t) b'(X_t) \, dt + b(X_t) \, dW_t \tag{9.5}$$

for a given differentiable function b is equivalent to the Stratonovich SDE

$$dX_t = b(X_t) \circ dW_t. \tag{9.6}$$

We can integrate the Stratonovich SDE (9.6) directly via

$$\frac{1}{b(X_t)} dX_t = 1 \circ dW_t = dW_t,$$

and obtain the general solution

$$X_t = h^{-1}(W_t + h(X_0))$$

where

$$y = h(x) = \int^x \frac{ds}{b(s)}.$$

In practice we would not use this explicit solution formula directly, but rather we would repeat the above procedure of first converting to the corresponding Stratonovich SDE and then integrating it with an appropriate substitution.

Example 9.1.2 The Ito SDE

$$dX_t = \frac{1}{2} a^2 m X_t^{2m-1} dt + a X_t^m dW_t, \qquad m \neq 1,$$

is equivalent to the Stratonovich SDE

$$dX_t = a X_t^m \circ dW_t$$

which we can directly integrate via

$$\frac{1}{X_t^m} dX_t = a \circ dW_t = a \, dW_t,$$

to obtain the solution

$$X_t = \left(X_0^{1-m} - a(m-1) W_t \right)^{1/(1-m)}.$$

Here $h(x) = x^{1-m}/(1-m)$.

Maple Assistance 83

Here we can use the MAPLE procedure *ItoS* defined in the previous chapter to convert the given Ito SDE to its corresponding Stratonovich SDE, and then use MAPLE to solve it.

> ItoS(1/2*a^2*m*x^(2*m-1),a*x^m);

$$dX = \left(\frac{1}{2} a^2 m X^{(2m-1)} - \frac{1}{2} \frac{a^2 (X^m)^2 m}{X} \right) dt + a X^m SdW$$

> simplify(%);

$$dX = a X^m SdW$$

That is,

> diff(X(W),W)=a*X(W)^m;

$$\frac{\partial}{\partial W} X(W) = a X(W)^m$$

> simplify(dsolve(%,X(W)));

$$X(W) = (a W - a W m + \_C1)^{\left(-\frac{1}{m-1} \right)}$$

Reducible Case 2

We now start with the Ito SDE

$$dX_t = \left(\alpha b(X_t) + \frac{1}{2} b(X_t) b'(X_t) \right) dt + b(X_t) dW_t, \qquad (9.7)$$

which is equivalent to the Stratonovich SDE

$$dX_t = \alpha\, b(X_t)\, dt + b(X_t) \circ dW_t. \tag{9.8}$$

Thus

$$\frac{1}{b(X_t)}\, dX_t = \alpha\, dt + 1 \circ dW_t = \alpha\, dt + 1\, dW_t,$$

which we integrate to obtain the solution

$$X_t = h^{-1}(\alpha t + W_t + h(X_0)).$$

Here h is the same as in case 1.

Example 9.1.3 The Ito SDE

$$dX_t = (1 + X_t)\left(1 + X_t^2\right)\, dt + \left(1 + X_t^2\right)\, dW_t$$

has equivalent Stratonovich SDE

$$dX_t = \left(1 + X_t^2\right)\, dt + \left(1 + X_t^2\right) \circ dW_t.$$

Integrating

$$\frac{1}{\left(1 + X_t^2\right)}\, dX_t = 1\, dt + 1 \circ dW_t = 1\, dt + 1\, dW_t,$$

we obtain the solution

$$X_t = \tan\left(t + W_t + \arctan X_0\right),$$

since $h(x) = \arctan(x)$ here.

MAPLE ASSISTANCE 84

We define a procedure reducible2 that solves Ito SDEs that are reducible (case 2).

```
>  reducible2:=proc(a,b)
>  local temp1,h,soln,W,alpha;
>  temp1:=alpha*b+1/2*b*simplify(diff(b,x));
>  temp1=a;
>  alpha:=simplify(solve(%,alpha));
>  if diff(alpha,x)<>0 then
>  ERROR('non-linear SDE not of this reducible type')
>  else
>  h:=int(1/b,x);
>  soln:=h=alpha*t+W+subs(x=X[0],h);
>  soln:=X[t]=simplify(solve(%,x));
>  fi;
>  end:
```

Next, we use this procedure to solve the given Ito SDE. Two input parameters are required; the drift a and the diffusion b. Note that input must be in terms of x.

```
>  reducible2((1+x)*(1+x^2),(1+x^2));
```
$$X_t = \tan(t + W + \arctan(X_0))$$

$$\cdots$$

There are many other possibilities, all reducing to a Stratonovich SDE with a separable structure as above in its x variable and leading to a stochastic differential with constant coefficients. Ito SDEs that are reducible to a linear Ito SDEs will be considered in Section 9.3.

9.2 Linear Scalar SDEs

As with linear ordinary differential equations, the general solution of a linear stochastic differential equation can be determined explicitly with the help of an integrating factor or, equivalently, a fundamental solution of an associated homogeneous differential equation.

The general form of a scalar *linear Ito SDE* is

$$dX_t = (a_1(t)X_t + a_2(t))\ dt + (b_1(t)X_t + b_2(t))\ dW_t \qquad (9.9)$$

where the coefficients a_1, a_2, b_1, b_2 are specified functions of time t or constants. Provided they are Lebesgue measurable and bounded on an interval $0 \le t \le T$, the existence and uniqueness theorem applies, ensuring the existence of a unique solution X_t on $t \in [t_0, T]$ for any $t_0 \in [0, T]$ and each $\mathcal{A}_{t_0}$-measurable initial value X_{t_0} corresponding to a given Wiener process $\{W_t, t \ge 0\}$ and its associated family of σ-algebras $\{\mathcal{A}_t, t \ge 0\}$.

When $a_2(t) \equiv 0$ and $b_2(t) \equiv 0$, (9.9) reduces to the *homogeneous bilinear Ito SDE*

$$dX_t = a_1(t)X_t\ dt + b_1(t)X_t\ dW_t, \qquad (9.10)$$

Obviously $X_t \equiv 0$ is a solution of (2.2), but the *fundamental solution* Φ_{t,t_0} with the initial condition $\Phi_{t_0,t_0} = 1$ is of far greater significance since all other solutions can be expressed in terms of it. The task is to find such a fundamental solution.

When $b_1(t) \equiv 0$ in (9.9) the SDE has the form

$$dX_t = (a_1(t)X_t + a_2(t))\ dt + b_2(t)\ dW_t, \qquad (9.11)$$

that is the noise appears *additively* rather than multiplicatively as in (9.10). Its homogeneous equation, obtained from (9.10) by setting $b_1(t) \equiv 0$, is the deterministic ordinary differential equation

$$\frac{dX_t}{dt} = a_1(t)X_t \tag{9.12}$$

for which the fundamental solution (in fact, the inverse of its integrating factor) is

$$\Phi_{t,t_0} = \exp\left(\int_{t_0}^{t} a_1(s)\, ds\right). \tag{9.13}$$

The Ito formula (8.21) applied to the transformation function $U(t,x) = \Phi_{t,t_0}^{-1} x$ (note that $\Phi_{t,t_0}^{-1} = \Phi_{t_0,t}$ here) and the solution X_t of (9.10) gives

$$d\left(\Phi_{t,t_0}^{-1} X_t\right) = \left(\frac{d\Phi_{t,t_0}^{-1}}{dt} X_t + (a_1(t)X_t + a_2(t))\, \Phi_{t,t_0}^{-1}\right)\, dt + b_2(t)\Phi_{t,t_0}^{-1}\, dW_t$$

$$= a_2(t)\Phi_{t,t_0}^{-1}\, dt + b_2(t)\Phi_{t,t_0}^{-1}\, dW_t,$$

since

$$\frac{d\Phi_{t,t_0}^{-1}}{dt} = -\Phi_{t,t_0}^{-1} a_1(t)$$

Note that the right hand side of

$$d\left(\Phi_{t,t_0}^{-1} X_t\right) = a_2(t)\Phi_{t,t_0}^{-1}\, dt + b_2(t)\Phi_{t,t_0}^{-1}\, dW_t$$

only involves known functions of t (and ω), so it can be integrated to give

$$\Phi_{t,t_0}^{-1} X_t = \Phi_{t_0,t_0}^{-1} X_{t_0} + \int_{t_0}^{t} a_2(s)\Phi_{s,t_0}^{-1}\, ds + \int_{t_0}^{t} b_2(s)\Phi_{s,t_0}^{-1}\, dW_s.$$

Since $\Phi_{t_0,t_0} = 1$ this leads to the solution

$$X_t = \Phi_{t,t_0}\left(X_{t_0} + \int_{t_0}^{t} a_2(s)\Phi_{s,t_0}^{-1}\, ds + \int_{t_0}^{t} b_2(s)\Phi_{s,t_0}^{-1}\, dW_s\right) \tag{9.14}$$

of the additive noise linear SDE (9.11), where the fundamental solution Φ_{t,t_0} is defined by (9.13).

Example 9.2.1 The Langevin equation

$$dX_t = -aX_t\, dt + dW_t \tag{9.15}$$

is a linear SDE with additive noise. The coefficients here are $a_1(t) \equiv -a$, $a_2(t) \equiv 0$, $b_1(t) \equiv 0$ and $b_2(t) \equiv b$. Its fundamental solution is $\Phi_{t,t_0} = \exp(-a(t-t_0))$ and the solution for $t_0 = 0$ is given by

$$X_t = X_0 e^{-at} + e^{-at}\int_{0}^{t} e^{as}\, dW_s.$$

The general linear case is more complicated because the associated homogeneous equation is a genuine stochastic differential equation. Its fundamental solution is now

$$\Phi_{t,t_0} = \exp\left(\int_{t_0}^t \left(a_1(s) - \frac{1}{2}b_1^2(s)\right)ds + \int_{t_0}^t b_1(s)\,dW_s\right), \qquad (9.16)$$

with $\Phi_{t_0,t_0} = 1$, and the resulting solution of (9.9) is

$$X_t = \Phi_{t,t_0}\left(X_{t_0} + \int_{t_0}^t (a_2(s) - b_1(s)b_2(s))\,\Phi_{s,t_0}^{-1}\,ds + \int_{t_0}^t b_2(s)\Phi_{s,t_0}^{-1}\,dW_s\right).$$
$$(9.17)$$

Example 9.2.2 The bilinear SDE

$$dX_t = aX_t\,dt + bX_t\,dW_t \qquad (9.18)$$

with coefficients $a_1(t) \equiv a$, $b_1(t) \equiv b$ and $a_2(t) \equiv b_2(t) \equiv 0$, has the fundamental solution

$$\Phi_{t,t_0} = \exp\left(\int_{t_0}^t \left(a - \frac{1}{2}b^2\right)(t - t_0) + b(W_t - W_{t_0})\right)$$

and the general solution

$$X_t = X_{t_0}\exp\left(\int_{t_0}^t \left(a - \frac{1}{2}b^2\right)(t - t_0) + b(W_t - W_{t_0})\right).$$

MAPLE ASSISTANCE 85

We define a procedure Linearsde that solves general linear Ito SDEs of the form (9.9).

```
> Linearsde:=proc(a,b)
> local temp1,alpha,beta,gamma,delta,fundsoln,fundsoln2,
> soln,default1,default2,default3;
> if diff(a,x,x)<>0 or diff(b,x,x)<>0 then
> ERROR('SDE not linear')
> else
> alpha:=diff(a,x);
> alpha:=subs(t=s,alpha);
> beta:=diff(b,x);
> beta:=subs(t=s,beta);
> if diff(beta,s)=0 then
> temp1:=beta*W;
> else temp1:=Int(beta,W=0..t);
> fi;
> gamma:=coeff(a,x,0);
```

```
> gamma:=subs(t=s,gamma);
> delta:=coeff(b,x,0);
> delta:=subs(t=s,delta);
> fundsoln:=exp(int(alpha-1/2*(beta^2),s=0..t)+temp1);
> fundsoln2:=subs(t=s,fundsoln);
> if beta=0 then
> soln:=fundsoln*(X[0]+int(1/fundsoln2*(gamma-beta*delta),
  s=0..t)
> +Int(1/fundsoln2*delta,W=0..t))
> else
> soln:=fundsoln*(X[0]+Int(1/fundsoln2*(gamma-beta*delta),
  s=0..t)
> +Int(1/fundsoln2*delta,W=0..t))
> fi;
> default1:=Int(0,W=0..t)=0;
> default2:=Int(0,W=0..s)=0;
> default3:=Int(0,s=0..t)=0;
> soln:=X[t]=subs(default1,default2,default3,soln)
> fi
> end:
```

Now we apply this procedure to the Langevin equation (9.15) and also to the bilinear SDE (9.18). There are two input parameters required; the drift a and the diffusion b. Again, input must be in terms of x.

```
> Linearsde(-a*x,1);
```

$$X_t = e^{(-at)} \left(X_0 + \int_0^t \frac{1}{e^{(-as)}} \, dW \right)$$

```
> Linearsde(a*x,b*x);
```

$$X_t = e^{(at-1/2\,b^2\,t+b\,W)} X_0$$

In this example the solution is given for $t_0 = 0$.

9.2.1 Moment Equations

If we take the expectation of the integral form of equation (9.9) and use the zero expectation property of an Ito integral, we obtain an ordinary differential equation for the expected value or mean $m(t) = \mathbb{E}(X_t)$ of its solution, namely

$$\frac{dm(t)}{dt} = a_1(t)m(t) + a_2(t). \tag{9.19}$$

If use the Ito formula to obtain an SDE for X_t^2 and then take the expectation of the integral form of this equation (See Example 8.5.1 in the previous chapter), we also find that the second order moment $P(t) = \mathbb{E}\left(X_t^2\right)$ satisfies the ordinary differential equation

$$\frac{dP(t)}{dt} = \left(2a_1(t) + b_1^2(t)\right) P(t) + 2m(t)\left(a_2(t) + b_1(t)b_2(t)\right) \qquad (9.20)$$

$$+b_2^2(t).$$

MAPLE ASSISTANCE 86

Below we present procedures moment1 and moment2, which automate construction of equations (9.19) and (9.20), respectively.

```
>   moment1:=proc(a1,a2)
>   diff(m(t),t)=a1*m(t)+a2;
>   end:
```

Note the input parameters required for procedure moment1 a1 and a2 correspond to $a_1(t)$ and $a_2(t)$ in equation (9.9).

```
>   moment2:=proc(a1,a2,b1,b2)
>   diff(p(t),t)=(2*a1+b1^2)*p(t)+2*m(t)*(a2+b1*b2)+b2^2;
>   end:
```

Also, in this procedure the input parameters correspond to the variables of equation (9.9). Sometimes it might be most convenient to solve the first moment equation for $m(t)$, substitute the solution to the second equation and then solve it for $P(t)$.

9.3 Scalar SDEs Reducible to Linear SDEs

With an appropriate substitution $X_t = U(t, Y_t)$ certain nonlinear stochastic differential equations

$$dY_t = a(t, Y_t)\, dt + b(t, Y_t)\, dW_t \qquad (9.21)$$

can be reduced to a linear SDE in X_t

$$dX_t = \left(a_1(t)X_t + a_2(t)\right) dt + \left(b_1(t)X_t + b_2(t)\right) dW_t. \qquad (9.22)$$

The Inverse Function Theorem ensures the existence of a local inverse $y = V(t, x)$ of $x = U(t, y)$, i.e. with $x = U(t, V(t, x))$ and $y = V(t, U(t, y))$, provided $\frac{\partial U}{\partial y}(t, y) \neq 0$. A solution Y_t of (9.21) then has the form $Y_t = V(t, X_t)$ where X_t is given by (9.22) for appropriate coefficients a_1, a_2, b_1 and b_2. From the Ito formula (8.21) we have

$$dU(t, Y_t) = \left(\frac{\partial U}{\partial t} + a\frac{\partial U}{\partial y} + \frac{1}{2}b^2\frac{\partial^2 U}{\partial y^2}\right) dt + b\frac{\partial U}{\partial y}\, dW_t,$$

where the coefficients and the partial derivatives are all evaluated at (t, Y_t). This coincides with a linear SDE of the form (9.22) when

$$\frac{\partial U}{\partial t}(t,y) + a(t,y)\frac{\partial U}{\partial y}(t,y) + \frac{1}{2}b^2(t,y)\frac{\partial^2 U}{\partial y^2}(t,y) \qquad (9.23)$$

$$= a_1(t)U(t,y) + a_2(t)$$

and

$$b(t,y)\frac{\partial U}{\partial y}(t,y) = b_1(t)U(t,y) + b_2(t). \qquad (9.24)$$

Little more can be said at this level of generality that is directly applicable, so we shall specialize to the case where $a_1(t) \equiv b_1(t) \equiv 0$ and writing $a_2(t) = \alpha(t)$ and $b_2(t) = \beta(t)$, i.e. where the linear SDE (9.22) is the stochastic differential

$$dX_t = \alpha(t)\,dt + \beta(t)\,dW_t. \qquad (9.25)$$

Then from (9.23) we obtain the identity

$$\frac{\partial^2 U}{\partial t \partial y}(t,y) = -\frac{\partial}{\partial y}\left(a(t,y)\frac{\partial U}{\partial y}(t,y) + \frac{1}{2}b^2(t,y)\frac{\partial^2 U}{\partial y^2}(t,y)\right)$$

and, from (9.24), the identities

$$\frac{\partial}{\partial y}\left(b(t,y)\frac{\partial U}{\partial y}(t,y)\right) = 0$$

and

$$b(t,y)\frac{\partial^2 U}{\partial t \partial y}(t,y) + \frac{\partial b}{\partial t}(t,y)\frac{\partial U}{\partial y}(t,y) = \beta'(t).$$

Supposing that $b(t,y) \neq 0$, we can eliminate U and its derivatives to obtain

$$\beta'(t) = \beta(t)b(t,y)\left(\frac{1}{b^2(t,y)}\frac{\partial b}{\partial t}(t,y) - \frac{\partial}{\partial y}\left(\frac{a(t,y)}{b(t,y)}\right) + \frac{1}{2}\frac{\partial^2 b}{\partial y^2}(t,y)\right).$$

Since the left hand side is independent of y this means

$$\frac{\partial \gamma}{\partial y}(t,y) = 0 \qquad (9.26)$$

where

$$\gamma(t,y) = \frac{1}{b(t,y)}\frac{\partial b}{\partial t}(t,y) - b(t,y)\frac{\partial}{\partial y}\left(\frac{a(t,y)}{b(t,y)} - \frac{1}{2}\frac{\partial b}{\partial y}(t,y)\right). \qquad (9.27)$$

This is a sufficient condition for the reducibility of the nonlinear SDE (9.21) to the explicitly integrable stochastic differential (9.25) by means of a transformation $x = U(t,y)$. In this special case the identities (9.23)–(9.24) reduce to

$$\frac{\partial U}{\partial t}(t,y) + a(t,y)\frac{\partial U}{\partial y}(t,y) + \frac{1}{2}b^2(t,y)\frac{\partial^2 U}{\partial y^2}(t,y) = \alpha(t) \tag{9.28}$$

and

$$b(t,y)\frac{\partial U}{\partial y}(t,y) = \beta(t), \tag{9.29}$$

which give

$$U(t,y) = C \exp\left(\int_0^t \gamma(s,y)\,ds\right)\int_0^y \frac{1}{b(t,z)}\,dz, \tag{9.30}$$

where C is an arbitrary constant.

Example 9.3.1 For the nonlinear SDE

$$dY_t = -\frac{1}{2}\exp\left(-2Y_t\right)dt + \exp\left(-Y_t\right)dW_t \tag{9.31}$$

$a(t,y) = -\frac{1}{2}\exp(-2y)$ and $b(t,y) = \exp(-y)$, so $\gamma(t,y) \equiv 0$. Thus with $U = \exp(y)$ we see that (9.31) reduces to the stochastic differential $dX_t = dW_t$ with $\alpha(t) \equiv 0$ and $\beta(t) \equiv 1$, which has solution

$$X_t = W_t + X_0 = W_t + \exp(Y_0).$$

Since $X_t = \exp(Y_t)$, the solution of the original nonlinear SDE (9.31) is thus

$$Y_t = \ln\left(W_t + \exp(Y_0)\right).$$

MAPLE ASSISTANCE 87

Below we give procedures for calculating γ, α and β defined above by (9.27)–(9.29) and then use these procedures to solve (9.31).

```
>   mgamma:=proc(a,b)
>   local part1,soln;
>   part1:=a/b-1/2*diff(b,y);
>   soln:=gamma=simplify(1/b*diff(b,t)-b*diff(part1,y));
>   end:
>   malpha:=proc(u,a,b)
>   alpha=diff(u,t)+a*diff(u,y)+1/2*b^2*diff(diff(u,y),y);
>   simplify(%);
>   end:
>   mbeta:=proc(u,b)
>   beta=b*diff(u,y);
>   simplify(%);
>   end:
```

Note that the input for the above procedures must be given in terms of y. Below we illustrate the usage of these procedures by applying them to (9.31). First, we calculate γ, α and β.

```
>   mgamma(-1/2*exp(-2*y),exp(-y));
```
$$\gamma = 0$$

```
>   malpha(exp(y),-1/2*exp(-2*y),exp(-y));
```
$$\alpha = 0$$

```
>   mbeta(exp(y),exp(-y));
```
$$\beta = 1$$

Now we can use the previously defined procedure `reducible2` to solve the resulting SDE $dX_t = \alpha dt + \beta dW_t$, which for this example is $dX_t = dW_t$.

```
>   reducible2(0,1);
```
$$X_t = W + X_0$$

```
>   subs(X[0]=exp(Y[0]),%);
```
$$X_t = W + e^{Y_0}$$

```
>   subs(X[t]=exp(Y[t]),%);
```
$$e^{Y_t} = W + e^{Y_0}$$

```
>   Y[t]=solve(%,Y[t]);
```
$$Y_t = \ln(W + e^{Y_0})$$

9.4 Vector SDES

Consider m scalar Wiener processes $W_t^1, W_t^2, \ldots, W_t^m$ with a common family of increasing σ-algebras $\{A_t, t \geq 0\}$ that are pairwise independent, i.e. each W_t^j is A_t-measurable with

$$\mathbb{E}\left(W_t^j | A_0\right) = 0, \qquad \mathbb{E}\left(W_t^j - W_s^j | A_s\right) = 0,$$

and

$$\mathbb{E}\left((W_t^i - W_s^i)(W_t^j - W_s^j)|A_s\right) = (t - s)\,\delta_{i,j},$$

w.p.1, for $0 \leq s \leq t$ and $i, j = 1, 2, \ldots, m$, where $\delta_{i,j}$ is the Kronecker delta symbol defined by

$$\delta_{i,j} = \begin{cases} 1 & i = j \\ 0 & \text{otherwise} \end{cases}.$$

We denote the m-dimensional column vector with W_t^j as its jth component by W_t and call it an m-dimensional Wiener process. Here and in what follows, we interpret a vector as a column vector and use superscripts to index its components.

Consider also the n-dimensional column vector functions $a, b^1, \ldots, b^m$: $[0, T] \times \mathbb{R}^n \longrightarrow \mathbb{R}^n$. Then we denote by

$$dX_t = a(t, X_t)\, dt + \sum_{j=1}^{m} b^j(t, X_t)\, dW_t^j, \qquad (9.32)$$

an *n-dimensional vector Ito SDE driven by an m-dimensional Wiener process*, which we interpret as a vector Ito stochastic integral equation

$$X_t = X_{t_0} + \int_{t_0}^{t} a(s, X_s)\, ds + \sum_{j=1}^{m} \int_{t_0}^{t} b^j(s, X_s)\, dW_s^j. \qquad (9.33)$$

The Lebesgue and Ito stochastic integrals here are determined component by component with the ith component of (9.32) being

$$X_t^i = X_{t_0}^i + \int_{t_0}^{t} a^i(s, X_s)\, ds + \sum_{j=1}^{m} \int_{t_0}^{t} b^{i,j}(s, X_s)\, dW_s^j$$

for $i = 1, \ldots, n$ and an analogous definition of a solution to the scalar case is used.

Remark 9.4.1 *The indexing on the $b^{i,j}$ here appears to be the reverse of what may have been expected, with this term representing the ith component of the column vector b^j. In fact, $b^{i,j}$ written in this way is the (i, j)th component of the nxm-matrix $B = [b^1 | \cdots | b^m]$ with b^j as its jth column vector. We could have written the vector SDE (9.32) as*

$$dX_t = a(t, X_t)\, dt + B(t, X_t)\, dW_t,$$

where W_t is the m-dimensional Wiener process with the pairwise independent scalar Wiener processes $W_t^1, \ldots, W_t^m$ as its components.

Vector SDEs arise naturally in systems described by vector valued states. They also occur when certain scalar equations are reformulated to have this form.

Example 9.4.1 A second order ordinary differential equation disturbed by white noise that describes the dynamics of a satellite is given by

$$\frac{d^2 x}{d^2 t} + b\,(1 + a\,\xi_t)\,\frac{dx}{dt} + (1 + a\,\xi_t)\sin x - c\sin 2x = 0 \qquad (9.34)$$

where x is the radial perturbation about the given orbit, ξ_t is a Gaussian white noise and a, b and c are constants, with b and c positive. With $X_t = (X_t^1, X_t^2)^\top$ where $X^1 = x$ and $X^2 = \frac{dx}{dt}$, this can be rewritten as a 2-dimensional Ito SDE

$$d\begin{pmatrix} X_t^1 \\ X_t^2 \end{pmatrix} = \begin{pmatrix} X_t^2 \\ -b\,X_t^2 - \sin X_t^1 - c\sin 2X_t^1 \end{pmatrix} dt \qquad (9.35)$$

$$+ \begin{pmatrix} 0 \\ -ab\,X_t^2 - b\sin X_t^1 \end{pmatrix} dW_t$$

where W_t is a scalar Wiener process.

9.4.1 Vector Ito Formula

For a sufficiently smooth vector functions $U : [0, T] \times \mathbb{R}^n \longrightarrow \mathbb{R}$ the scalar transformed process $Y_t = U(t, X_t)$ of a solution X_t of the vector SDE (9.32) has the stochastic differential

$$dY_t = L^0 U(t, X_t)\, dt + \sum_{j=1}^{m} L^j U(t, X_t)\, dW_t^l, \tag{9.36}$$

where the operators L^0, L^1, ..., L^m are defined by

$$L^0 = \frac{\partial}{\partial t} + \sum_{i=1}^{n} a^i \frac{\partial}{\partial x_i} + \frac{1}{2} \sum_{i,j=1}^{n} \sum_{l=1}^{m} b^{i,l} b^{j,l} \frac{\partial^2}{\partial x_i \partial x_j} \tag{9.37}$$

and

$$L^j = \sum_{i=1}^{n} b^{i,j} \frac{\partial}{\partial x_i} \tag{9.38}$$

for $j = 1, ..., m$. This is general form of the *Ito formula* for vector SDEs.

Example 9.4.2 Let X_t^1 and X_t^2 satisfy the scalar stochastic differentials

$$dX_t^i = e_t^i\, dt + f_t^i\, dW_t^i \tag{9.39}$$

for $i = 1, 2$, where the coefficient functions are nonanticipative random functions. Let $U(t, x_1, x_2) = x_1 x_2$. Then the stochastic differential satisfied by the product process

$$Y_t = X_t^1 X_t^2$$

depends on whether the Wiener processes W_t^1 and W_t^2 are independent or dependent. In the former case the differentials (9.39) can be written as the vector differential

$$d \begin{pmatrix} X_t^1 \\ X_t^2 \end{pmatrix} = \begin{pmatrix} e_t^1 \\ e_t^2 \end{pmatrix} dt + \begin{bmatrix} f_t^1 & 0 \\ 0 & f_t^2 \end{bmatrix} d \begin{pmatrix} W_t^1 \\ W_t^2 \end{pmatrix} \tag{9.40}$$

and the transformed differential is

$$dY_t = \left(e_t^1 X_t^2 + e_t^2 X_t^1 \right) dt + f_t^1 X_t^2\, dW_t^1 + f_t^2 X_t^1\, dW_t^2. \tag{9.41}$$

In contrast, when $W_t^1 = W_t^2 = W_t$ the vector differential for (9.39) is

$$d \begin{pmatrix} X_t^1 \\ X_t^2 \end{pmatrix} = \begin{pmatrix} e_t^1 \\ e_t^2 \end{pmatrix} dt + \begin{pmatrix} f_t^1 \\ f_t^2 \end{pmatrix} dW_t \tag{9.42}$$

and there is an extra term $f_t^1 f_t^2\, dt$ in the differential of Y_t, which is now

$$dY_t = \left(e_t^1 X_t^2 + e_t^2 X_t^1 + f_t^1 f_t^2\right) dt + \left(f_t^1 X_t^2 + f_t^2 X_t^1\right) dW_t. \tag{9.43}$$

In the special case that $e_t^1 \equiv e_t^2 \equiv 0$ and $f_t^1 \equiv f_t^2 \equiv 1$, we have $X_t^1 = W_t^1$, $X_t^2 = W_t^2$ and the stochastic differential

$$d\left(X_t^1 X_t^2\right) = d\left(W_t^1 W_t^2\right) = W_t^2 \, dW_t^1 + W_t^1 \, dW_t^2 \tag{9.44}$$

in the first case, and $X_t^1 = X_t^2 = W_t$ and the stochastic differential

$$d\left(X_t^1 X_t^2\right) = d\left((W_t)^2\right) = 1 \, dt + 2W_t \, dW_t \tag{9.45}$$

in the second case. Similarly, if we set $e_t^1 \equiv 1$, $e_t^2 \equiv 0$, $f_t^1 \equiv 0$ and $f_t^2 \equiv 1$ in (9.42), so $X_t^1 = t$ and $X_t^2 = W_t$, then we obtain the stochastic differential

$$d\left(X_t^1 X_t^2\right) = d\left(t W_t\right) = W_t \, dt + t \, dW_t. \tag{9.46}$$

These three expressions, and similar ones, will be used in the next chapter when we present higher order numerical schemes for SDEs.

MAPLE ASSISTANCE 88

Before we give the MAPLE procedure for the general form of the Ito formula for vector SDEs defined by equation (9.36) we will present procedures L0 and LJ, which automate application of the operators defined by equations (9.37) and (9.38), respectively.

```
>   L0:=proc(X::algebraic,a::list(algebraic),
    b::list(list(a lgebraic)))
>   local part1,part2,part3;
>   part1:=diff(X,t);
>   part2:=sum('a[k]*diff(X,x[k])','k'=1..nops(a));
>   part3:=1/2*sum('sum('sum('op(j,op(k,b))
>   *op(j,op(l,b))*diff(X,x[k],x[l])','j'=1..nops(op(1,b)))',
>   'k'=1..nops(a))','l'=1..nops(a));
>   part1+part2+part3
>   end:
>   LJ
    := proc(X::algebraic,b::list(list(algebraic)),j::integer)
>   sum('op(j,op(k,b))*diff(X,x[k])','k'=1..nops(b))
>   end:
```

Now procedure Ito, which may be applied to any vector Ito SDE, is given below.

```
>   Ito := proc(X::algebraic,a::list(algebraic),
    b::list(list( algebraic)))
>   dY = L0(X,a,b)*dt+sum('LJ(X,b,j)*dW.j',
    'j'=1..nops(op(1,b )));
>   end:
```

Next, procedure Ito is applied to the two vector Ito SDEs (9.40) and (9.42) above to obtain the transformed differentials given by equations (9.41) and (9.43), respectively.

```
> Ito(x[1]*x[2],[e[1],e[2]],[[f[1],0],[0,f[2]]]);
```
$$dY = (e_1 x_2 + e_2 x_1)\, dt + f_1 x_2\, dW1 + f_2 x_1\, dW2$$

```
> Ito(x[1]*x[2],[e[1],e[2]],[[f[1]],[f[2]]]);
```
$$dY = (e_1 x_2 + e_2 x_1 + f_2 f_1)\, dt + (f_1 x_2 + f_2 x_1)\, dW1$$

Note the input required for procedure Ito; the drift a must be given as an array of terms, and the diffusion b as an array of arrays where the ith array contains the diffusion components for the ith equation of the vector Ito SDE. All input variables must be in terms of $x[i]$.

The output contains Wiener processes W_t^j denoted in MAPLE by Wj.

9.4.2 Fokker-Planck Equation

The transition probabilities of the vector Ito SDE (9.32) have transition probability densities $p(s, x; t, y)$ which satisfy the *Fokker-Planck equation*,

$$\frac{\partial p}{\partial t} + \sum_{i=1}^{n} \frac{\partial}{\partial y_i} \left\{ a^i(t,y)p \right\} - \frac{1}{2} \sum_{i,j=1}^{n} \sum_{k=1}^{m} \frac{\partial^2}{\partial y_i \partial y_j} \left\{ b^{i,k}(t,y) b^{j,k}(t,y)p \right\} = 0.$$

$$(9.47)$$

MAPLE ASSISTANCE 89

Below we present a MAPLE procedure that constructs equation (9.47). Procedure FPE below requires the input parameters a and b to be of the same form as the corresponding input parameters a and b in procedure L0 defined in the previous section.

```
> FPE:=proc(a,b)
> local soln,part1,part2,part3;
> part1:=Diff(p,t);
> part2:=sum('Diff(a[i],x[i])*p','i'=1..nops(a));
> part3:=-1/2*sum('sum('sum('Diff(op(k,op(i,b))*
    op(k,op(j,b))*p,x[i],x[j])',
> 'k'=1..nops(op(1,b)))','i'=1..nops(a))','j'=1..nops(a)) ;
> soln:=part1+part2+part3=0;
> subs(seq(x[i]=y[i],i=1..nops(a)),soln);
> end:
```

9.5 Vector Linear SDE

The general form of an n-dimensional *linear stochastic differential equation* is

$$dX_t = (A(t)X_t + a(t)) \, dt + \sum_{j=1}^{m} \left(B^j(t)X_t + b^j(t)\right) \, dW_t^j \qquad (9.48)$$

where $A(t)$, $B^1(t)$, $B^2(t)$, ..., $B^m(t)$ are $n{\times}n$-matrix functions and $a(t)$, $b^1(t)$, $b^2(t)$, ..., $b^m(t)$ are n-dimensional vector functions. When a and the $b^1(t)$, ...,$b^m(t)$ are all identically zero we obtain the *homogeneous equation*.

$$dX_t = A(t)X_t \, dt + \sum_{j=1}^{m} B^j(t)X_t \, dW_t^j \qquad (9.49)$$

and when the B^1, B^2, ..., B^m are all identically zero we have the *linear SDE with additive noise*

$$dX_t = (A(t)X_t + a(t)) \, dt + \sum_{j=1}^{m} b^j(t) \, dW_t^j \qquad (9.50)$$

Analogously to the argument used for the scalar case. it can be shown that the solution of (9.48) is

$$X_t = \Phi_{t,t_0} \left(X_{t_0} + \int_{t_0}^{t} \Phi_{s,t_0}^{-1} \left(a(s) - \sum_{j=1}^{m} B^j(s)b^j(s) \right) \, ds \right.$$

$$\left. + \sum_{j=1}^{m} \int_{t_0}^{t} \Phi_{s,t_0}^{-1} b^j(s) \, dW_s^j \right), \qquad (9.51)$$

where Φ_{t,t_0} is the $n \times n$ fundamental matrix satisfying $\Phi_{t_0,t_0} = I$ and the homogeneous *matrix SDE*

$$d\Phi_{t,t_0} = A(t)\Phi_{t,t_0} \, dt + \sum_{j=1}^{m} B^j(t)\Phi_{t,t_0} \, dW_t^j, \qquad (9.52)$$

which is interpreted column vector by column vector as vector stochastic differential equations.

Unlike the scalar homogeneous linear equations, we cannot generally solve (9.52) explicitly for its fundamental solution, even when all of the matrices are constant matrices. If, however, the matrices A, B^1, B^2, ..., B^m are constants and commute, i.e. if

$$AB^l = B^l A \quad \text{and} \quad B^l B^k = B^k B^l$$

for all $k, l = 1, 2, \ldots, m$, then we obtain the following explicit expression for the fundamental matrix solution

$$\Phi_{t,t_0} = \exp\left(\left(A - \frac{1}{2}\sum_{j=1}^{m}(B^j)^2\right)(t - t_0) + \sum_{j=1}^{m} B^j\left(W_t^j - W_{t_0}^j\right)\right). \quad (9.53)$$

In the special case that (9.48) has additive noise, i.e. (9.50), then (9.53) reduces to

$$\Phi_{t,t_0} = \exp\left((t - t_0)A\right),$$

which is the fundamental matrix of the deterministic linear system

$$\dot{x} = Ax.$$

In such autonomous cases $\Phi_{t,t_0} = \Phi_{t-t_0,0}$, so we need only to consider $t_0 = 0$ and could use Φ_t for $\Phi_{t,0}$.

Example 9.5.1 Solve the random oscillator equation

$$dX_t = \begin{bmatrix} 0 & -1 \\ 1 & 0 \end{bmatrix} X_t\, dt + \begin{pmatrix} 0 \\ \sigma^2 \end{pmatrix} dW_t,$$

where $X_t = (X_t^1, X_t^2)$ and W_t is a scalar Wiener process, by first showing that the fundamental matrix is

$$\Phi_t = \begin{bmatrix} \cos t & -\sin t \\ \sin t & \cos t \end{bmatrix}.$$

MAPLE ASSISTANCE 90

Since the noise is additive here, the fundamental matrix is the same as that of a deterministic linear system, with

$$A = \begin{bmatrix} 0 & -1 \\ 1 & 0 \end{bmatrix}$$

Now, we can construct the appropriate fundamental matrix using the command **exponential**, which is in the **linalg** package of MAPLE.

```
>  with(linalg):
>  a:=matrix(2,2,[0,-1,1,0]):
>  exponential(%,t);
```

$$\begin{bmatrix} \cos(t) & -\sin(t) \\ \sin(t) & \cos(t) \end{bmatrix}$$

9.5.1 Moment Equations

In the same way as for scalar linear SDEs, we can derive vector and matrix ordinary differential equations for the vector mean $m(t) = \mathbb{E}(X_t)$ and the $n \times n$ matrix second moment $P(t) = \mathbb{E}(X_t X_t^\mathsf{T})$ of a general vector linear SDE (9.48). Recall that for n-dimensional column vectors x and y, the product xy^T is a $n \times n$ matrix with (i, j)th component $x^i y^j$. Thus $P(t)$ is a symmetric matrix. We then obtain

$$\frac{dm}{dt} = A(t)m + a(t) \tag{9.54}$$

and

$$\frac{dP}{dt} = A(t)P + PA(t)^\mathsf{T} + \sum_{l=1}^{m} B^l(t)PB^l(t)^\mathsf{T} \tag{9.55}$$

$$+ a(t)m(t)^\mathsf{T} + m(t)a(t)^\mathsf{T}$$

$$+ \sum_{l=1}^{m} \left(B^l(t)m(t)b^l(t)^\mathsf{T} + b^l(t)m(t)^\mathsf{T}B^l(t) + b^l(t)b^l(t)^\mathsf{T} \right),$$

with initial conditions $m(t_0) = \mathbb{E}(X_{t_0})$ and $P(t_0) = \mathbb{E}(X_{t_0} X_{t_0}^\mathsf{T})$.

Note, that the both above equations are linear, the equation for $m(t)$ does not depend on $P(t)$, so can be solved first, and then its solution can be substituted into the second equation.

9.5.2 Linearization

We consider the linearization of an n-dimensional Ito SDE (9.32) about a fixed solution $\bar{X}_t$. This results in the n-dimensional linear vector SDE

$$dZ_t = A(t)Z_t\, dt + \sum_{j=1}^{m} B^j(t)Z_t\, dW_t^j \tag{9.56}$$

where

$$A(t)^{i,j} = \frac{\partial a^i}{\partial x^j}(t, \bar{X}(t)), \qquad B^k(t)^{i,j} = \frac{\partial b^{k,i}}{\partial x^j}(t, \bar{X}(t))$$

for $i,\, j = 1,\, \ldots,\, n$ and $k = 1,\, \ldots,\, m$. These coefficient functions will be constants when the solution $\bar{X}_t$ is a constant such as 0. Otherwise they will be random functions, i.e. depending on ω through the corresponding sample path of $\bar{X}_t$, but these random functions will be nonanticipative so the stochastic integral in the integral equation version of (9.56) will be meaningful.

MAPLE ASSISTANCE 91

The MAPLE routine below, linearize, automates the linearizing of vector Ito SDEs about a stationary solution $\bar{X}_t$

```
> linearize:=proc(a,b,c)
> local i,tempA,tempB,j,k,l;
> tempA:=array(1..nops(a),1..nops(a));
> for i from 1 to nops(a) do for j from 1 to nops(a) do
> tempA[i,j]:=diff(op(i,a),x[j]);
> od; od;
> for i from 1 to nops(a) do
> for j from 1 to nops(a) do
> for l from 1 to nops(c) do
> tempA[i,j]:=subs(x[l]=op(l,c),tempA[i,j]);
> od; od; od;
> for k from 1 to nops(op(1,b)) do
> tempB[k]:=array(1..nops(a),1..nops(a));
> for i from 1 to nops(a) do
> for j from 1 to nops(a) do
> tempB[k][i,j]:=diff(op(k,op(i,b)),x[j]);
> od; od;
> for i from 1 to nops(a) do
> for j from 1 to nops(a) do
> for l from 1 to nops(c) do
> tempB[k][i,j]:=subs(x[l]=op(l,c),tempB[k][i,j]);
> od; od; od;
> od;
> RETURN(A=map(simplify,convert(eval(tempA),matrix)),
    B=eval(tempB))
> end:
```

The input required for procedure linearize are the variables $A(t)$ and $B^j(t)$ from equation (9.56), and the stationary solution $\bar{X}_t$. The first input parameter must be given as an array, the second as an array of arrays, and the third as an array. Input variables must be in the form $x[j]$.

9.6 Vector Stratonovich SDEs

There are vector analogues of the relationships between vector Ito and Stratonovich stochastic differential equations. In particular, the solution X_t of the vector Ito SDE

$$dX_t = a(t, X_t)\, dt + \sum_{j=1}^{m} b^j(t, X_t)\, dW_t^j \tag{9.57}$$

is also a solution of the equivalent vector Stratonovich SDE

$$dX_t = \underline{a}(t, X_t)\, dt + \sum_{j=1}^{m} b^j(t, X_t) \circ dW_t^j \tag{9.58}$$

with the modified drift

$$\underline{a}(t, x) = a(t, x) - c(t, x), \tag{9.59}$$

where the n-dimensional vector function $c = c(t, X)$ is defined componentwise by

$$c^i(t, x) = \frac{1}{2} \sum_{j=1}^{n} \sum_{k=1}^{m} b^{j,k}(t, x) \frac{\partial b^{i,k}}{\partial x_j}(t, x) \tag{9.60}$$

for $i = 1, 2, \ldots, n$.

MAPLE ASSISTANCE 92

A MAPLE routine is presented below that automates the conversion process between Ito calculus and Stratonovich calculus for vector SDEs.

```
>   conv:=proc(a,b,c)
>   local temp,i;
>   if c=ito then
>   for i from 1 to nops(a) do
>   temp[i]:=op(i,a)-1/2*sum('sum('op(k,op(j,b))*
    diff(op(k,op(i,b)),x[j])', 'k'=1..nops(op(1,b)))',
    'j'=1..nops(a));
>   od;
>   elif c=strat then
>   for i from 1 to nops(a) do
>   temp[i]:=op(i,a)+1/2*sum('sum('op(k,op(op(j,b))*
    diff(op(k,op(i,b)),x[j])',
>   'k'=1..nops(op(1,b)))','j'=1..nops(a));
>   od;
>   else
>   ERROR(
    'Must enter either ito or strat for the 3rd argument')
>   fi;
>   RETURN(map(simplify,eval(temp)))
>   end:
```

The first two input parameters for procedure conv are the drift and diffusion. Drift must be entered as an array, and diffusion as an array of arrays. Input variables must be of the form $x[i]$. The third input parameter is used to tell conv what type of SDE, either Ito or Stratonovich, we wish to convert. For example, entering "strat" would indicate a request to convert a Stratonovich SDE to an Ito SDE.

. . .

Example 9.6.1 The drift correction term $c(t, x) \equiv 0$ is possible even when the diffusion coefficients depend explicitly on x. Consider, for example, a 2-dimensional SDE involving a scalar Wiener process with diffusion coefficient $b = (b^1, b^2)$ satisfying

$$b^1(x_1, x_2) = -b^2(x_1, x_2) = x_1 + x_2.$$

9.7 MAPLE Session

M 9.1 *Define a MAPLE procedure, called* `explicit`, *which combines the previously defined procedures* `Linearsde` *and* `reducible2`. *The procedure should determine whether a given SDE is reducible (case 2) or linear, and solve it accordingly.*

A suitable procedure is given below.
```
>  explicit:=proc(a,b)
>  if diff(a,x,x)=0 and diff(b,x,x)=0 then
>  Linearsde(a,b)
>  else
>  reducible2(a,b)
>  fi
>  end:
```
Procedure `explicit` is more user-friendly than `Linearsde` or `reducible2` as it is not necessary for the user to predetermine which category the SDE belongs to.

M 9.2 *Write a procedure to simulate paths of the standard Wiener process. Then, plot a single path.*

```
>  with(stats,random):
>  W_path := proc(T,n)
   local w, h, t, alist:
   w := 0: t := 0: h := T/n: alist := [0,w]:
   from 1 to n do
   t := t + h:
   w := w + random[normald[0,sqrt(h)]](1):
   alist := alist,[t,w]:
   od:
   [alist]:
   end:
>  W_path(2,100):
>  plots[pointplot](%);
```

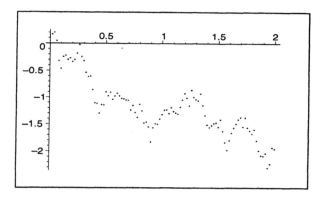

M 9.3 *Plot a solution path of the Ito equation from Example 9.1.3, assuming* $X_0 = 1$.

We have just find the formula for the solution, see page 252.

```
>  X := tw -> [op(1,tw),evalf(tan( op(1,tw) + op(2,tw) ))];
```
$$X := tw \to [op(1,\, tw),\, \text{evalf}(\tan(op(1,\, tw) + op(2,\, tw)))]$$
```
>  map(X,W_path(2,100)):
>  plots[pointplot](%);
```

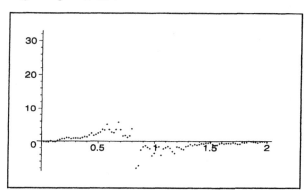

M 9.4 *Plot a solution path of the bilinear SDE (9.18), assuming* $a = 1$, $b = 2$, $X_0 = 30$.

We have just find the formula for the solution, see page 256.

```
>  a := 1: b := 2:
>  X := proc(tw)
     local t,w:
     t := op(1,tw):
     w := op(2,tw):
```

```
      [t,30*exp(a*t - b^2/2*t+b*w)]
      end:
>   map(X,W_path(5,200)):
>   plots[pointplot](%);
```

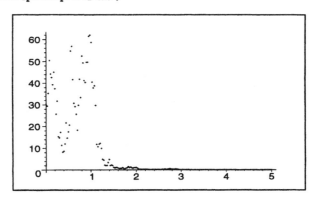

M 9.5 *Find the first and second moments for the equation*

$$dX_t = (-.05\,X_t + 10)\;dt + (0.1\,X_t + 0.5)\;dW_t$$

with $X_0 \equiv const = 100$. *Plot confidence intervals for the mean of the solution.*

We use procedures (9.19) and (9.20).
```
>   moment1(-.5,100);
```
$$\frac{\partial}{\partial t}\,m(t) = -.5\,m(t) + 100$$
```
>   moment2(-.5,10,0.1,0.5);
```
$$\frac{\partial}{\partial t}\,p(t) = -.99\,p(t) + 20.10\,m(t) + .25$$
```
>   dsolve({%,%%, m(0) = 100, p(0) = 0},{m(t),p(t)});
```

$$\{p(t) = \frac{402025}{99} - \frac{201000}{49}\,e^{(-1/2\,t)} + \frac{199775}{4851}\,e^{(-\frac{99}{100}\,t)},$$
$$m(t) = 200 - 100\,e^{(-1/2\,t)}\}$$
```
>   assign(%);
>   m := unapply(m(t),t); p := unapply(p(t),t);
```
$$m := t \to 200 - 100\,e^{(-1/2\,t)}$$
$$p := t \to \frac{402025}{99} - \frac{201000}{49}\,e^{(-1/2\,t)} + \frac{199775}{4851}\,e^{(-\frac{99}{100}\,t)}$$
```
>   std := t -> p(t) - m(t)^2;
```
$$std := t \to p(t) - m(t)^2$$

Assuming that X_t are normally distributed (the validity of which we will discuss later) we plot symmetric 0.95-confidence intervals.

```
>   plot([m - 1.96*sqrt@p,m, m + 1.96*sqrt@p], 0..20,
    linestyle= [3,1,3],color = BLACK);
```

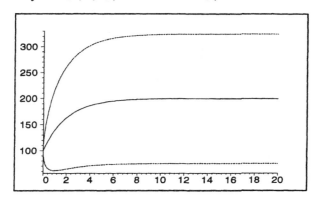

M 9.6 *Consider the following 2-dimensional Stratonovich SDE*

$$dX_t^1 = X_t^2 dt$$
$$dX_t^2 = (-bX_t^2 - \sin X_t^1 - c\sin 2X_t^1 - a^2(X_t^2)^3 + aX_t^2 \sin(X_t^1))dt$$
$$+ (-a(X_t^2)^2 + \sin(X_t^1)) \circ dW_t \tag{9.61}$$

where a, b and c are constants and W_t is a scalar Wiener process. Use the previously defined procedure conv to convert the above Stratonovich SDE to a system of Ito SDEs.

```
>   answer := conv([x[2],-b*x[2]-sin(x[1])-c*sin(2*x[1])
    -a^2*x[2]^3+a*x[2]*sin(x[1])],[[0],
    [-a*(x[2])^2+sin(x[1])]], strat);
```

$$answer := \text{table}([$$
$$1 = x_2$$
$$2 = -bx_2 - \sin(x_1) - c\sin(2x_1)$$
$$])$$

The elements of the above table represent the corrected drift matrix.

M 9.7 *Use the drift matrix obtained in the previous exercise, defined by the variable* **answer**, *to linearize the resulting system of Ito SDEs about the steady solution* $(\bar{X}_t^1, \bar{X}_t^2) \equiv (0,0)$.

Below we use the previously defined procedure `linearize` to achieve the result.

```
> linearize([answer[1],answer[2]],[[0],
  [-a*(x[2])^2+sin(x [1])]],[0,0]);
```

$$A = \begin{bmatrix} 0 & 1 \\ -1 - 2\,c & -b \end{bmatrix}, B = \text{table}([$$

$$1 = \begin{bmatrix} 0 & 0 \\ 1 & 0 \end{bmatrix}$$

])

M 9.8 *Determine the time dependent expectation vector for the solution of 2-dimensional system:*

$$dX_t = (A(t)X_t + a(t))\, dt + \sum_{j=1}^{m} \left(B^j(t)X_t + b^j(t) \right)\, dW_t^j, \quad X_0 = \begin{bmatrix} 2 \\ 3 \end{bmatrix}$$

with

$$A(t) \equiv \begin{bmatrix} -1 & 3 \\ -3 & -1 \end{bmatrix}, \quad a(t) \equiv \begin{bmatrix} -5 \\ 7 \end{bmatrix}.$$

We will find the solution of the first moment equation (9.54) using MAPLE `matrixDE` procedure.

```
> A := matrix(2,2,[-1,3,-3,-1]);
```

$$A := \begin{bmatrix} -1 & 3 \\ -3 & -1 \end{bmatrix}$$

```
> a := matrix(2,1,[-5,7]);
```

$$a := \begin{bmatrix} -5 \\ 7 \end{bmatrix}$$

```
> msol := DEtools[matrixDE](A,a,t);
```

$$msol := \begin{bmatrix} \begin{bmatrix} e^{(-t)} \sin(3\,t) & e^{(-t)} \cos(3\,t) \\ e^{(-t)} \cos(3\,t) & -e^{(-t)} \sin(3\,t) \end{bmatrix}, \begin{bmatrix} \dfrac{8}{5} & \dfrac{11}{5} \end{bmatrix} \end{bmatrix}$$

```
> constants := matrix(2,1,[C1,C2]);
```

$$constants := \begin{bmatrix} C1 \\ C2 \end{bmatrix}$$

```
>  gsol := evalm(msol[1]&*constants + msol[2]);
```

$$gsol := \begin{bmatrix} e^{(-t)} \sin(3\,t)\,C1 + e^{(-t)} \cos(3\,t)\,C2 + \dfrac{8}{5} \\[2ex] e^{(-t)} \cos(3\,t)\,C1 - e^{(-t)} \sin(3\,t)\,C2 + \dfrac{11}{5} \end{bmatrix}$$

```
>  m1 := unapply(gsol[1,1],t);
```

$$m1 := t \to e^{(-t)} \sin(3\,t)\,C1 + e^{(-t)} \cos(3\,t)\,C2 + \frac{8}{5}$$

```
>  m2 := unapply(gsol[2,1],t);
```

$$m2 := t \to e^{(-t)} \cos(3\,t)\,C1 - e^{(-t)} \sin(3\,t)\,C2 + \frac{11}{5}$$

```
>  solve( {m1(0) = 2, m2(0) = 3}, {C1,C2} );
```

$$\{C1 = \frac{4}{5},\ C2 = \frac{2}{5}\}$$

```
>  assign(%);
```

Now, we can plot the mean.

```
>  plot([m1(t),m2(t),t= 0..5]);
```

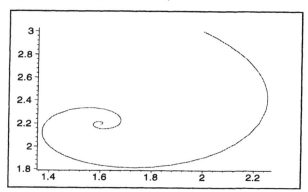

9.8 Exercises

E 9.1 Determine confidence intervals for the solution mean for the Langevin equation (9.15).

E 9.2 Generate 100 solution paths for the bilinear equation with specific parameters, initial condition and time interval length. Then check graphically the normality of X_t for a few values of t.

E 9.3 Determine and solve the ordinary differential equations for the first and second moments of the 2-dimensional linear SDE

$$dX_t = -2X_t\,dt + \begin{bmatrix} 0 & -1 \\ 1 & 0 \end{bmatrix} X_t\,dW_t,$$

where $X_t = (X_t^1, X_t^2)^\top$ and W_t is a scalar Wiener process.

E 9.4 Show that the Ito SDE

$$dX_t = -\beta^2 X_t \left(1 - X_t^2\right) dt + \beta \left(1 - X_t^2\right) dW_t$$

has the solution

$$X_t = \frac{(1 + X_0) \exp(2\beta W_t) + X_0 - 1}{(1 + X_0) \exp(\beta W_t) + 1 - X_0}.$$

E 9.5 Solve the scalar SDE

$$dX_t = -\frac{1}{2} X_t \, dt + X_t \, dW_t^1 + X_t \, dW_t^2,$$

where W_t^1 and W_t^2 are independent scalar Wiener processes.

E 9.6 Determine the vector Stratonovich SDE corresponding to the vector Ito SDE

$$dX_t = -\frac{1}{2} X_t \, dt + \begin{bmatrix} 0 & -1 \\ 1 & 0 \end{bmatrix} X_t \, dW_t,$$

where W_t is a scalar Wiener process.

E 9.7 Linearize the 2-dimensional Stratonovich SDE (9.61) in the MAPLE session exercise **MS** 9.6 about $X^1 = X^2 = 0$, then convert the linearized Stratonovich SDE into an Ito SDE and compare the result with the Ito SDE obtained in the MAPLE session exercise **MS** 9.7.

E 9.8 Determine the vector Stratonovich SDE corresponding to the vector Ito SDE

$$dX_t = -\frac{1}{2} X_t \, dt + \begin{bmatrix} 0 & -1 \\ 1 & 0 \end{bmatrix} X_t \, dW_t,$$

where W_t is a scalar Wiener process.

10 Numerical Methods for SDEs

To begin we briefly recall some background material on the numerical solution of deterministic ordinary differential equations (ODEs) to provide an introduction to the corresponding ideas for SDEs. We then discuss the stochastic Euler scheme and its implementation in some detail before turning to higher order numerical schemes for scalar Ito SDEs. We also indicate how MAPLE can be used to automatically determine the coefficients of these numerical schemes, in particular for vector SDEs.

10.1 Numerical Methods for ODEs

The simplest numerical scheme for the ordinary differential equation

$$\frac{dx}{dt} = a(t, x), \tag{10.1}$$

with initial condition $x(t_0) = x_0$ is the *Euler scheme* with constant time step $t_{n+1} - t_n =: \Delta > 0$,

$$x_{n+1} = x_n + a(t_n, x_n)\,\Delta, \qquad n = 0, 1, 2, \ldots. \tag{10.2}$$

It is an example of an 1-step explicit scheme, which have the general form

$$x_{n+1} = x_n + F(\Delta, t_n, x_n)\,\Delta, \qquad n = 0, 1, 2, \ldots, \tag{10.3}$$

where $F(\Delta, t, x)$ is the *increment function* of the scheme. For the Euler scheme, $F(\Delta, t, x) = a(t, x)$. It is the order $N = 1$ truncated Taylor scheme for (10.1) and can be derived by truncating the first order Taylor expansion of the solution of (10.1). The general order $N \geq 1$ truncated Taylor schemes for a scalar ODE (10.1) is given by

$$x_{n+1} = x_n + \Delta \sum_{j=1}^{N} \frac{1}{j!} D^{j-1} a(t_n, x_n)\,\Delta^{j-1}, \qquad n = 0, 1, 2, \ldots, \tag{10.4}$$

where D is the *total derivative* defined by

$$D = \frac{\partial}{\partial t} + a\,\frac{\partial}{\partial x}$$

with $D^0 a = a$. Its increment function is thus

$$F(\Delta, t, x) = \sum_{j=1}^{N} \frac{1}{j!} D^{j-1} a(t, x) \Delta^{j-1}.$$

The error in truncating a deterministic Taylor expansion after N terms is proportional to Δ^{N+1}, where Δ is the length of the time interval under consideration. This gives the estimate

$$|x_{n+1} - x(t_{n+1}; t_n, x_n)| \leq C \Delta^{N+1} \tag{10.5}$$

for the *local discretization error* of the Nth order truncated Taylor scheme (10.4), where $x(t; t_n, x_n)$ is the solution of the ODE (10.1) starting from x_n at time t_n. However, this is not the error that we want since we need to compare the numerical iterates x_n with the values $x(t_n; t_0, x_0)$ of the solution $x(t; t_0, x_0)$ of (10.1) with initial value $x(t_0) = x_0$. A classical result of numerical analysis says that we lose an order in going from an estimate of the local discretization error to one for the *global discretization error*, which is given by

$$|x_n - x(t_n; t_0, x_0)| \leq K_T \Delta^N, \tag{10.6}$$

where the constant K_T depends amongst other things on the length T of the time interval $[t_0, t_0 + T]$ under consideration and the power N is the order of the scheme rather than $N + 1$ in (10.5). The deterministic Euler scheme (10.2) thus has order $N = 1$.

Higher order truncated Taylor schemes have been rarely used in practice because the $D^{j-1} a(t, x)$ coefficients can often be very complicated and unwieldy for $j \geq 3$, although MAPLE now mitigates this difficulty. Instead they were (and still are) typically used theoretically to determine the order of other schemes for which the increment function $F(\Delta, t, x)$ has been obtained by a variety of heuristic methods. (An important criterion for the convergence of such schemes is their *consistency*, i.e. $F(0, t, x) \equiv a(t, x)$, although this does not say anything about the rate or order of convergence). The derivative-free Runge–Kutta schemes are amongst the most commonly implemented schemes of the form (10.3). In practice, 4th order Runga-Kutta schemes are usually used, but here we consider the simpler *Heun scheme*

$$x_{n+1} = x_n + \frac{1}{2} \left[a(t_n, x_n) + a(t_{n+1}, x_n + a(t_n, x_n)\Delta) \right] \Delta, \tag{10.7}$$

which is of 2nd order and has the increment function

$$F(\Delta, t, x) = \frac{1}{2} \left[a(t, x) + a(t + \Delta, x + a(t, x)\Delta) \right]. \tag{10.8}$$

It is derived essentially by averaging the values of a at both ends of the discretization interval $[t_n, t_{n+1}]$, i.e. $a(t_n, x_n)$ and $a(t_{n+1}, x_{n+1})$, and then replacing the implicit value x_{n+1} at the upper endpoint by its Euler approximation $x_n + a(t_n, x_n)\Delta$.

Example 10.1.1 To show that the Heun scheme (10.7) is of 2nd order, expand the second term in the increment function (10.8) about (t, x) to get

$$a(t + \Delta, x + a(t,x)\Delta) = a + (a_t + aa_x)\,\Delta + \frac{1}{2}\left(a_{tt} + 2aa_x + a^2 a_{xx}\right)\Delta^2 + \cdots,$$

where all of the functions on the righthand side are evaluated at (t, x), and so

$$x_{n+1} = x_n + a(t_n, x_n)\,\Delta + \frac{1}{2}Da(t_n, x_n)\,\Delta^2 + O(\Delta^3).$$

The Heun scheme (10.7) thus coincides with the order $N = 2$ truncated Taylor scheme up to the remainder term, so also has local discretization error of order 3 and hence global discretization error of order 2, see [4].

We will compare the numerical solutions obtained by the Euler and Heun schemes with the "true solution" of the ODE

$$\frac{dx}{dt} = x - t^2 \text{ with initial condition } x(0) = 1.$$

MAPLE ASSISTANCE 93

We define the right hand side of the equation:
```
>   a := (t,x) -> x - t^2 ;
```
$$a := (t, x) \to x - t^2$$

We find the "true solution":
```
>   dsolve( {diff(x(t),t) = a(t,x(t)),x(0) = 1},x(t) );
```
$$x(t) = t^2 + 2t + 2 - e^t$$

```
>   true_sol := unapply(rhs(%),t);
```
$$true\_sol := t \to t^2 + 2t + 2 - e^t$$

```
>   p1 := plot(true_sol,0..3):
```

We compute the numerical solution according to the Euler scheme, see (10.2) and (10.3).
```
>   F := a:
>   t := 0: x := 1: Delta := 0.1: alist := [t,x]:
>   for i from 1 to 30 do
    x := x + F(t,x)*Delta: t := t + Delta:
    alist := alist,[t,x]:
    od:
>   p2 := plots[pointplot]([ alist ]):
```

Now compare "true" and numerical solution.
```
>   plots[display](p1,p2,axes=FRAMED);
```

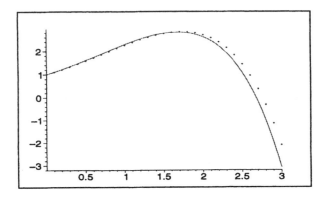

Now, do the same using the Heun scheme, see (10.8).

```
>  F := (Delta,t,x) ->
   1/2*(a(t,x)+ a(t+Delta,x+a(t,x)*Delta) );
```

$$F := (\varDelta, t, x) \to \frac{1}{2}\,\mathrm{a}(t, x) + \frac{1}{2}\,\mathrm{a}(t + \varDelta, x + \mathrm{a}(t, x)\,\varDelta)$$

```
>  t := 0: x := 1: Delta := 0.1: alist := [t,x]:
>  for i from 1 to 30 do
   x := x + F(Delta,t,x)*Delta: t := t + Delta:
   alist := alist,[t,x]:
   od:
>  p3 := plots[pointplot]([ alist ]):
>  plots[display](p1,p3,axes = FRAMED);
```

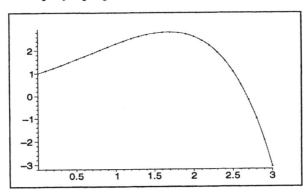

10.2 The Stochastic Euler Scheme

Deterministic calculus is much more robust than stochastic calculus because the integrand in a Riemann sum approximating a Riemann integral can be evaluated at an arbitrary point of the discretization subinterval, whereas for an Ito stochastic integral it must always be evaluated at the lower endpoint.

Consequently some care is required in deriving numerical schemes for Ito SDEs to ensure that they are consistent with Ito calculus. The stochastic counterpart of the Euler scheme (10.2) for a scalar Ito SDE

$$dX_t = a(t, X_t)\, dt + b(t, X_t)\, dW_t, \tag{10.9}$$

is given by

$$X_{n+1} = X_n + a(t_n, X_n)\, \Delta_n + b(t_n, X_n)\, \Delta W_n, \tag{10.10}$$

where

$$\Delta_n = t_{n+1} - t_n = \int_{t_n}^{t_{n+1}} ds, \qquad \Delta W_n = W_{t_{n+1}} - W_{t_n} = \int_{t_n}^{t_{n+1}} dW_s.$$

It appears to be consistent (and indeed is) with Ito stochastic calculus because the noise term in (10.10) approximates the Ito stochastic integral in (the integral version of) (10.9) over a discretization subinterval $[t_n, t_{n+1}]$ by evaluating its integrand at the lower end point of this interval, i.e.

$$\int_{t_n}^{t_{n+1}} b(s, X_s)\, dW_s \approx \int_{t_n}^{t_{n+1}} b(t_n, X_{t_n})\, dW_s$$

$$= b(t_n, X(t_n)) \int_{t_n}^{t_{n+1}} dW_s.$$

In contrast with the deterministic case, there are several different types of convergences that are useful for stochastic numerical schemes. We will distinguish between strong and weak convergence depending on whether we require the realisations or only their probability distributions to be close. We say that a stochastic numerical scheme converges with *strong order* γ if

$$\mathbb{E}\left(|X_T - X_{N_T}^{\Delta}|\right) \leq K_T\, \Delta^{\gamma} \tag{10.11}$$

and with with *weak order* β if

$$\left|\mathbb{E}(g(X_T)) - \mathbb{E}\left(g(X_{N_T}^{\Delta})\right)\right| \leq K_{g,T}\, \Delta^{\beta} \tag{10.12}$$

for each polynomial g, where $\Delta = \max_{0 \leq n \leq N_T - 1} \Delta_n$. The errors in (10.11) and (10.12) are global discretization errors, so γ and β are the corresponding strong and weak orders, respectively, of the scheme.

To implement the stochastic Euler scheme (10.10) note that the noise increment ΔW_n is an $N(0, \sqrt{\Delta_n})$ distributed Gaussian random variable, that is with first and second moments

$$\mathbb{E}(\Delta W_n) = 0, \qquad \mathbb{E}\left((\Delta W_n)^2\right) = \Delta_n,$$

and the ΔW_n are independent random variables for $n = 0, 1, 2, \ldots$. Realisations of such random variables can be done with MAPLE.

Example 10.2.1 We will construct a procedure which uses the Euler scheme (10.10) for plotting the solutions of a scalar Ito SDE,

MAPLE ASSISTANCE 94

```
>  with(stats,random):
>  EulerScheme := proc(a,b,X0,h,n)
   local x, t, alist:
   x := X0: t := 0: alist := [0,x]:
   from 1 to n do
   x := evalf(x + a(t,x)*h +
   b(t,x)*random[normald[0,sqrt(h)]](1)):
   t := t + h:
   alist := alist,[t,x]:
   od:
   [alist]:
   end:
```

Now we use the above procedure for the Langevin equation (9.15) and the bilinear equation (9.18).

```
>  EulerScheme((t,x) ->-0.5*x,(t,x)->1, 10,0.01,500):
   plots[pointplot](%);
```

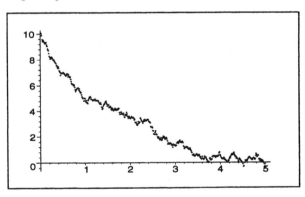

```
>  EulerScheme((t,x) -> x,(t,x)-> 1.5*x, 100,0.01,500):
>  plots[pointplot](%);
```

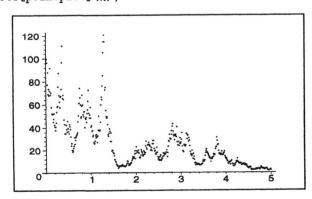

Remark 10.2.1 *Weak convergence ensures that all moments converge at the same rate although possibly with a different constant of proportionality $K_{g,T}$. Since*

$$\left| \mathbb{E}\left(X_T\right) - \mathbb{E}\left(X_{N_T}^\Delta\right) \right| \leq \mathbb{E}\left(\left|X_T - X_{N_T}^\Delta\right|\right),$$

we see that weak convergence with $g(x) \equiv x$ is implied by strong convergence.

A stochastic numerical scheme can, and typically does, have a different order of strong and weak convergence. For example, the stochastic Euler scheme (10.10) converges with strong order $\gamma = \frac{1}{2}$ and weak order $\beta = 1$. As with deterministic schemes, the order here is with respect to the general class of coefficient functions under consideration and a higher order may be possible for a more restricted class. (see Exercise **E** 10.2).

When weak convergence is of interest, the Gaussian increments ΔW_n in the stochastic Euler scheme (10.10) can be replaced by other more convenient approximations $\Delta \widehat{W}_n$ which have similar lower order moment properties to the ΔW_n, such as the two-point distributed random variable $\Delta \widehat{W}_n$ taking values $\pm\sqrt{\Delta_n}$ with equal probabilities, i.e. with

$$P\left(\Delta \widehat{W}_n = \pm\sqrt{\Delta_n}\right) = \frac{1}{2}. \tag{10.13}$$

This gives the *simplified stochastic Euler scheme*

$$X_{n+1} = X_n + a(t_n, X_n)\,\Delta_n + b(t_n, X_n)\,\Delta \widehat{W}_n, \tag{10.14}$$

which also has the weak order of convergence $\beta = 1$.

Realisations $\widehat{W}_n(\omega)$ of a two-point random variable (10.13) needed for implementation of the simplified stochastic Euler scheme (10.14) can be easily simulated with MAPLE. It is important in the case of a large scale simulations as the simulation of $\widehat{W}_n(\omega)$ is essentially faster than simulation of $W_n(\omega)$.

10.2.1 Statistical Estimates and Confidence Intervals

Often the goal of a simulation is to approximate an expectation

$$\mu = \mathbb{E}\left(f(X_T)\right) \tag{10.15}$$

of the solution of the stochastic differential equation (10.9) at time $T > t_0$ for some given function f by the expectation

$$\mu_\Delta = \mathbb{E}\left(f\left(X_{N_T}^\Delta\right)\right) \tag{10.16}$$

where $X_{N_T}^\Delta$ is the corresponding iterate of the numerical scheme with constant step size Δ, thus $N_T = (T - t_0)/\Delta$. In practice, (10.16) can only be estimated by the arithmetic mean

$$\hat{\mu}_{\Delta,M} = \frac{1}{M} \sum_{j=1}^{M} f\left(X_{N_T}^{\Delta}(\omega_j)\right) \tag{10.17}$$

of a <u>finite</u> number M of realisations of the numerical scheme. The estimate $\hat{\mu}_{\Delta,M}$ is itself the realisation of a random variable $\mu_{\Delta,M}$, depending on which M realisations of the numerical scheme are used to calculate it. To obtain a statistically reliable result confidence intervals should be used, with the other realisations of $\mu_{\Delta,M}$ lying within a $100(1-\alpha)\%$ confidence interval of the form $[\hat{\mu}_{\Delta,M} - \hat{\epsilon}_{\Delta,M,\alpha}, \hat{\mu}_{\Delta,M} + \hat{\epsilon}_{\Delta,M,\alpha}]$ with probability $1-\alpha$, where $\hat{\epsilon}_{\Delta,M,\alpha}$ is proportional to $1/\sqrt{M}$.

Applying the approach form Section 6.7 to the random variable $\xi = f\left(X_{N_T}^{\Delta}\right)$, recall that confidence intervals are determined with L batches of K realisations each, where $M = KL$ and $K \geq 20$. First, we calculate the arithmetic mean of each batch

$$\hat{\mu}_{\Delta,l} = \frac{1}{K} \sum_{k=1}^{K} f\left(X_{N_T}^{\Delta}(\omega_{k,l})\right)$$

and then their mean (which is the total sample mean)

$$\hat{\mu}_{\Delta,M} = \frac{1}{L} \sum_{l=1}^{L} \hat{\mu}_{\Delta,l} = \frac{1}{LK} \sum_{l=1}^{L} \sum_{k=1}^{K} f\left(X_{N_T}^{\Delta}(\omega_{k,l})\right).$$

The half-length $\hat{\epsilon}_{\Delta,M,\alpha}$ of the confidence interval about $\mu_{\Delta,M}$ is then given by

$$\hat{\epsilon}_{\Delta,M,\alpha} == \sqrt{\frac{\hat{\sigma}_l^2}{L}} t_{L-1}^{-1}\left(1 - \frac{\alpha}{2}\right),$$

where $\hat{\sigma}_l^2$ is the unbiased sample variance of the batch means, i.e.

$$\hat{\sigma}_l^2 = \frac{1}{L-1} \sum_{l=1}^{L} \left(\hat{\mu}_{\Delta,l} - \hat{\mu}_{\Delta,M}\right)^2$$

and $t_{1-\alpha,L-1}$ is the value of the Student t-distribution with $L-1$ degrees of freedom and confidence parameter $0 < \alpha < 1$, if L is small. For large L ($L \geq 30$ is large enough), we can use the normal distribution instead of t-distribution.

The expectation μ_Δ (10.16) is itself only an approximation of the desired value μ (10.15). How good an approximation it is will also depend on the accuracy of the stochastic numerical scheme.

10.3 How to Derive Higher Order Schemes

The orders of strong and weak convergence of the stochastic Euler scheme are dissapointingly low, given the fact that a large number of realisations usually need to be generated for most practical applications. Heuristic adaptations of well known deterministic numerical schemes should be avoided because they are often inconsistent with Ito calculus or, when they do converge to the corresponding solution of the Ito SDE they do not improve the order of convergence.

The deterministic Heun scheme (10.7) adapted to the Ito SDE (10.9) has the form

$$X_{n+1} = X_n$$

$$+ \frac{1}{2} \left[a(t_n, X_n) + a(t_{n+1}, X_n + a(t_n, X_n)\Delta_n + b(t_n, X_n)\Delta W_n) \right] \Delta_n,$$

$$+ \frac{1}{2} \left[b(t_n, X_n) + b(t_{n+1}, X_n + a(t_n, X_n)\Delta_n + b(t_n, X_n)\Delta W_n) \right] \Delta W_n.$$

It is not consistent with Ito calculus for SDEs with multiplicative noise, i.e. for which the diffusion coefficient $b(t, x)$ depends explicitly on x.

Example 10.3.1 For the Ito SDE $dX_t = X_t \, dW_t$ the adapted Heun scheme (10.18) with constant stepsize $\Delta_n = \Delta$ simplifies to

$$X_{n+1} = X_n + \frac{1}{2} X_n \left(2 + \Delta W_n \right) \Delta W_n$$

so $X_{n+1} - X_n = X_n \left(1 + \frac{1}{2}\Delta W_n \right) \Delta W_n$. The conditional expectation

$$\mathbb{E}\left(\frac{X_{n+1} - X_n}{\Delta} \,\middle|\, X_n = x \right) = \frac{x}{\Delta} \mathbb{E}\left(\Delta W_n + \frac{1}{2}(\Delta W_n)^2 \right)$$

$$= \frac{x}{\Delta} \left(0 + \frac{1}{2}\Delta \right) = \frac{1}{2} x$$

should approximate the drift term $a(t, x) \equiv 0$ of the SDE. In particular, the numerical solution generated by this scheme does not converge to the corresponding solution of the Ito SDE.

Even when such an adapted scheme converges, it need not attain higher order than the Euler scheme if it only uses the noise increments ΔW_n. Essentially, the simple increments $\Delta W_n = W_{t_{n+1}} - W_{t_n}$ do not provide enough information about the Wiener process W_t *inside* the discretization subinterval $[t_n, t_{n+1}]$ to ensure a higher order of approximation. The necessary information is contained within multiple stochastic integrals of W_t that occur in an appropriate stochastic Taylor expansion. Such expansions provide the basis for the higher order schemes that will be presented in the following sections, but first we will state some of their properties for multiple stochastic integrals and also introduce a convenient compact notation that will be useful in the schemes that follow.

10.3.1 Multiple Stochastic Integrals

Stochastic Taylor expansions involve multiple deterministic, stochastic and mixed multiple integrals such as

$$\int_{t_n}^{t_{n+1}} \int_{t_n}^{t} ds\, dt, \qquad \int_{t_n}^{t_{n+1}} \int_{t_n}^{t} dW_s\, dt,$$

$$\int_{t_n}^{t_{n+1}} \int_{t_n}^{t} dW_s\, dW_t, \qquad \int_{t_n}^{t_{n+1}} \int_{t_n}^{t} \int_{t_n}^{s} dW_r\, dW_s\, dW_t,$$

in addition to the single integrals or increments

$$\Delta_n = \int_{t_n}^{t_{n+1}} dt, \qquad \Delta W_n = \int_{t_n}^{t_{n+1}} dW_t.$$

In some simple cases we can evaluate these multiple integrals directly, e.g.

$$\int_{t_n}^{t_{n+1}} \int_{t_n}^{t} ds\, dt = \frac{1}{2}\Delta_n^2 \tag{10.18}$$

and for some others we can use the Ito formula (see Example 9.4.2) to obtain expressions such as

$$\int_{t_n}^{t_{n+1}} \int_{t_n}^{t} dW_s\, dW_t, = \frac{1}{2}\left\{(\Delta W_n)^2 - \Delta_n\right\} \tag{10.19}$$

$$\int_{t_n}^{t_{n+1}} \int_{t_n}^{t} \int_{t_n}^{r} dW_r\, dW_s\, dW_t = \frac{1}{3!}\left\{(\Delta W_n)^3 - \Delta W_n \Delta_n\right\}. \tag{10.20}$$

In general, there are no such nice formulas, but often the probability distribution of a multiple stochastic integral can be determined and used to simulate the integral. For example, the random variable

$$\Delta Z_n = \int_{t_n}^{t_{n+1}} \int_{t_n}^{t} dW_s\, dt, \tag{10.21}$$

is normally distributed with mean, variance and correlation with ΔW_n given by

$$\mathbb{E}(\Delta Z_n) = 0, \quad \mathbb{E}\left((\Delta Z_n)^2\right) = \frac{1}{3}\Delta_n^3, \quad \mathbb{E}(\Delta W_n\, \Delta Z_n) = \frac{1}{2}\Delta_n^2, \tag{10.22}$$

respectively. There is no difficulty in generating the pair of correlated normally distributed random variables ΔW_n, ΔZ_n using the transformation

$$\Delta W_n = V_n^1 \Delta_n^{1/2}, \qquad \Delta Z_n = \frac{1}{2}\left(V_n^1 + \frac{1}{\sqrt{3}}V_n^2\right)\Delta_n^{3/2}, \tag{10.23}$$

where V_n^1 and V_n^2 are independent $N(0, 1)$ distributed random variables that can be themselves generated by the Box–Muller method (for example).

The probability distribution for (10.21) can be verified by properties of Ito stochastic integrals and the identity

$$\int_{t_n}^{t_{n+1}} \int_{t_n}^{t} dW_s\, dt + \int_{t_n}^{t_{n+1}} \int_{t_n}^{t} ds\, dW_t = \int_{t_n}^{t_{n+1}} dW_t \int_{t_n}^{t_{n+1}} dt, \quad (10.24)$$

which is derived by an application of the Ito formula (see (9.46)).

Finally, we introduce a compact notation that will enable us to avoid cumbersome expressions for multiple stochastic integrals when vector Wiener processes are used. Let W_t^1, W_t^2, ..., W_t^m be m pairwise independent scalar Wiener processes and define

$$W_t^0 \equiv t.$$

Thus, in particular $dW_t^0 = dt$ and a "stochastic" integration with respect to W_t^0 will in fact be just a deterministic integration, i.e.

$$\int_{t_n}^{t_{n+1}} dW_t^0 = \int_{t_n}^{t_{n+1}} dt$$

and similarly within multiple integrals. Then we introduce a *multi-index* $(i_1, \ldots, i_m)$ of length m with components $i_j \in \{0, 1, \ldots, m\}$, which we use to label m-fold multiple integrals with the integrators dW^{i_j} appearing from left to right in the order given in the multi–index. For example, with $m = 1, 2, 3$ we have

$$I_{(i_1); t_n, t_{n+1}} = \int_{t_n}^{t_{n+1}} dW_s^{i_1}, \quad (10.25)$$

$$I_{(i_1, i_2); t_n, t_{n+1}} = \int_{t_n}^{t_{n+1}} \int_{t_n}^{s_2} dW_{s_1}^{i_1}\, dW_{s_2}^{i_2}, \quad (10.26)$$

$$I_{(i_1, i_2); t_n, t_{n+1}} = \int_{t_n}^{t_{n+1}} \int_{t_n}^{s_3} \int_{t_n}^{s_2} dW_{s_1}^{i_1}\, dW_{s_2}^{i_2}\, dW_{s_3}^{i_3}, \quad (10.27)$$

recalling that $dW_t^0 = dt$, so for example

$$I_{(1,0); t_n, t_{n+1}} = \int_{t_n}^{t_{n+1}} \int_{t_n}^{t} dW_s^1\, dW_t^0 = \int_{t_n}^{t_{n+1}} \int_{t_n}^{t} dW_s^1\, dt.$$

Similar identities to (10.24) hold for multiple stochastic integrals, such as

$$I_{(i_1, i_2); t_n, t_{n+1}} + I_{(i_2, i_1); t_n, t_{n+1}} = I_{(i_1); t_n, t_{n+1}}\, I_{(i_2); t_n, t_{n+1}} = \Delta W_n^{i_1} \Delta W_n^{i_2} \quad (10.28)$$

for i_1, $i_2 \in \{1, \ldots, m\}$ (we could also include 0 here). These can be verified by the Ito formula (see (9.45) in this case) and are very useful in simplifying numerical schemes for SDE with special structural properties, which is crucial because there is no elementary way to simulate multiple integrals such as

$$I_{(1,2);t_n,t_{n+1}} = \int_{t_n}^{t_{n+1}} \int_{t_n}^{t} dW_s^1 \, dW_t^2$$

when W_t^1 and W_t^2 are independent Wiener processes.

10.4 Higher Order Strong Schemes

If we add the $I_{(1,1);t_n,t_{n+1}}$ term in the stochastic Taylor expansion (8.28) to the stochastic Euler scheme (10.10) for the scalar Ito SDE (10.9), i.e. to the SDE

$$dX_t = a(t, X_t) \, dt + b(t, X_t) \, dW_t,$$

we obtain the *Milstein scheme*

$$X_{n+1} = X_n + a(t_n, X_n) \, \Delta_n + b(t_n, X_n) \, \Delta W_n + L^1 b(t_n, X_n) I_{(1,1);t_n,t_{n+1}},$$

which we can expand out as

$$X_{n+1} = X_n + a(t_n, X_n) \, \Delta_n + b(t_n, X_n) \, \Delta W_n$$

$$+ \frac{1}{2} b(t_n, X_n) \frac{\partial b}{\partial x}(t_n, X_n) \left\{ (\Delta W_n)^2 - \Delta_n \right\} \quad (10.29)$$

using the formula (10.19) for the multiple integral and the definition (8.23) for the operator L^1 to get $L^1 b = b\frac{\partial b}{\partial x}$.

Under appropriate conditions on the coefficients of the SDE the Milstein scheme (10.29) converges with strong order $\gamma = 1.0$, which is an improvement on the strong order $\gamma = \frac{1}{2}$ of the stochastic Euler scheme (10.10). For weak convergence both the Euler and Milstein schemes have weak order $\beta = 1.0$, so there is no improvement in convergence rate here. Note that the new term in the Milstein scheme vanishes when the SDE has additive noise since $\frac{\partial b}{\partial x}(t, x) \equiv 0$. For such SDEs the stochastic Euler and Milstein schemes are identical, so the stochastic Euler scheme converges with strong order $\gamma = 1.0$ for SDEs with additive noise.

By adding to the Milstein scheme the multiple integral terms of a stochastic Taylor expansion with the multi-indices $(1,0)$, $(0,1)$, $(0,0)$ and $(1,1,1)$, we obtain the order $\frac{3}{2}$ *strong Taylor scheme*

$$X_{n+1} = X_n + a(t_n, X_n) \, \Delta_n + b(t_n, X_n) \, \Delta W_n + L^1 b(t_n, X_n) I_{(1,1);t_n,t_{n+1}}$$

$$+ L^1 a(t_n, X_n) I_{(1,0);t_n,t_{n+1}} + L^0 b(t_n, X_n) I_{(0,1);t_n,t_{n+1}}$$

$$+ L^0 a(t_n, X_n) I_{(0,0);t_n,t_{n+1}} + L^1 L^1 b(t_n, X_n) I_{(1,1,1);t_n,t_{n+1}}$$

which has strong order $\gamma = \frac{3}{2}$. This expands out as

$$X_{n+1} = X_n + a\,\Delta_n + b\,\Delta W_n + \frac{1}{2}bb'\left\{(\Delta W_n)^2 - \Delta_n\right\} + b\frac{\partial a}{\partial x}\Delta Z_n$$

$$+ \left\{\frac{\partial b}{\partial t} + a\frac{\partial b}{\partial x} + \frac{1}{2}b^2\frac{\partial^2 b}{\partial x^2}\right\}\{\Delta W_n\,\Delta_n - \Delta Z_n\}$$

$$+ \frac{1}{2}\left\{\frac{\partial a}{\partial t} + a\frac{\partial a}{\partial x} + \frac{1}{2}b^2\frac{\partial^2 a}{\partial x^2}\right\}\Delta_n^2 \tag{10.30}$$

$$+ \frac{1}{2}b\left\{b\frac{\partial^2 b}{\partial x^2} + \left(\frac{\partial b}{\partial x}\right)^2\right\}\left\{\frac{1}{3}(\Delta W_n)^2 - \Delta_n\right\}\Delta W_n$$

where all of the functions on the right hand side are evaluated at (t_n, X_n). Here we have used the expressions (10.18), (10.21), (10.19) and (10.20) for the multiple integrals, the identity (10.24) for the integrals $\Delta Z_n = I_{(1,0);t_n,t_{n+1}}$ and $I_{(0,1);t_n,t_{n+1}}$, and the definitions (8.22) and (8.23) for the operators L^0 and L^1.

Strong Taylor schemes of even higher order can be derived in much the same way, with the strong orders γ increasing by $\frac{1}{2}$ for each successive scheme, starting with the stochastic Euler scheme (10.10) with strong order $\gamma = \frac{1}{2}$, the Milstein scheme (10.29) with strong order $\gamma = 1$ and the above order $\frac{3}{2}$ strong Taylor scheme (10.30) with strong order $\gamma = \frac{3}{2}$.

A practical disadvantage of higher order strong schemes based on stochastic Taylor expansions is that the derivatives of various orders of the drift and diffusion coefficients must be determined and then evaluated at each step in addition to the coefficients themselves. There are time discrete approximations which avoid the use of derivatives, which we will call Runge–Kutta schemes in analogy with similar schemes for ODEs. However, we emphasize that it is not always possible to use heuristic adaptations of deterministic Runge-Kutta schemes such as the Heun scheme for SDEs because of the difference between deterministic and stochastic calculi. For example, the *stochastic Runge–Kutta scheme* of strong order $\gamma = 1.0$

$$X_{n+1} = X_n + a(t_n, X_n)\,\Delta_n + b(t_n, X_n)\,\Delta_n W_n \tag{10.31}$$

$$+ \frac{1}{2\sqrt{\Delta_n}}\left\{b\left(\widehat{\Upsilon}_n\right) - b(t_n, X_n)\right\}\left\{(\Delta W_n)^2 - \Delta_n\right\}$$

with the supporting value

$$\widehat{\Upsilon}_n = X_n + b(t_n, X_n)\sqrt{\Delta_n}$$

can be obtained heuristically from the Milstein scheme (10.29) simply by replacing the coefficient $L^1 b(t, x)$ there by the difference approximation

$$\frac{1}{\sqrt{\Delta}}\left\{b\left(t, x + a(t, x)\,\Delta_n + b(t, x)\sqrt{\Delta_n}\right) - b(t, x)\right\}.$$

10.5 Higher Order Weak Schemes

Higher order weak Taylor schemes are also derived from appropriate stochastic Taylor expansions, but unlike their strong counterparts the orders of weak convergence are integers $\beta = 1, 2, \ldots$. As we saw earlier, the stochastic Euler scheme (10.10) is the order 1 weak Taylor scheme. If we add all of the double multiple stochastic integrals from the Ito-Taylor expansion to the stochastic Euler scheme, we obtain the *order 2 weak Taylor scheme*

$$X_{n+1} = X_n + a(t_n, X_n)\, \Delta_n + b(t_n, X_n)\, \Delta W_n$$

$$+ L^1 b(t_n, X_n) I_{(1,1);t_n,t_{n+1}} + L^1 a(t_n, X_n) I_{(1,0);t_n,t_{n+1}}$$

$$+ L^0 b(t_n, X_n) I_{(0,1);t_n,t_{n+1}} + L^0 a(t_n, X_n) I_{(0,0);t_n,t_{n+1}}$$

for the scalar Ito SDE (10.9). This expands out as

$$X_{n+1} = X_n + a\, \Delta_n + b\, \Delta W_n + \frac{1}{2} b \frac{\partial b}{\partial x} \left\{ (\Delta W_n)^2 - \Delta_n \right\}$$

$$+ \left\{ \frac{\partial b}{\partial t} + a \frac{\partial b}{\partial x} + \frac{1}{2} b^2 \frac{\partial^2 b}{\partial x^2} \right\} \{\Delta W_n \Delta_n - \Delta Z_n\} \quad (10.32)$$

$$+ b \frac{\partial a}{\partial x} \Delta Z_n + \frac{1}{2} \left\{ \frac{\partial a}{\partial t} + a \frac{\partial a}{\partial x} + \frac{1}{2} b^2 \frac{\partial^2 a}{\partial x^2} \right\} \Delta_n^2,$$

where all of the functions on the right hand side are evaluated at (t_n, X_n) and $\Delta Z_n = I_{(1,0);t_n,t_{n+1}}$. Note that it is the same as the order $\frac{3}{2}$ strong Taylor scheme (10.30) without the final term with multi-index $(1, 1, 1)$. This is a typical feature of weak Taylor schemes.

As with weak convergence for the stochastic Euler scheme, we can use simpler random variables to approximate ΔW_n and ΔZ_n provided their moments coincide up to the order of accuracy under consideration. In this case we can replace ΔW_n and ΔZ_n in (10.32) by

$$\Delta \widehat{W}_n = \Delta_n^{1/2} T_n \qquad \text{and} \qquad \Delta \widehat{Z}_n = \frac{1}{2} \Delta_n^{3/2} T_n \qquad (10.33)$$

where the T_n are independent three-point distributed random variables with

$$P\left(T_n = \pm\sqrt{3}\right) = \frac{1}{6} \quad \text{and} \quad P\left(T_n = 0\right) = \frac{2}{3}. \qquad (10.34)$$

This gives us the *simplified order 2.0 weak Taylor scheme*

$$X_{n+1} = X_n + a\, \Delta_n + b\, \Delta_n^{1/2} T_n + \frac{1}{2} bb' \left(T_n^2 - 1\right) \Delta_n$$

$$+ \frac{1}{2} \left\{ b \frac{\partial a}{\partial x} + \frac{\partial b}{\partial t} + a \frac{\partial b}{\partial x} + \frac{1}{2} b^2 \frac{\partial^2 b}{\partial x^2} \right\} \Delta_n^{3/2} T_n \qquad (10.35)$$

$$+ \frac{1}{2} \left\{ \frac{\partial a}{\partial t} + a \frac{\partial a}{\partial x} + \frac{1}{2} b^2 \frac{\partial^2 a}{\partial x^2} \right\} \Delta_n^2,$$

where all of the functions on the right hand side are evaluated at (t_n, X_n)

10.6 The Euler and Milstein Schemes for Vector SDEs

We now consider an n-dimensional Ito SDE with m–dimensional Wiener process W_t with scalar components $W_t^1, W_t^2, \ldots, W_t^m$, which we write in vector form as

$$dX_t = a(t, X_t)\, dt + \sum_{j=1}^{m} b^j(t, X_t)\, dW_t^j,$$

and componentwise as

$$dX_t^i = a^i(t, X_t)\, dt + \sum_{j=1}^{m} b^{i,j}(t, X_t)\, dW_t^j,$$

for $i = 1, \ldots, n$, noting as before that $b^{i,j}$ denotes the i-th component of the vector b^j. The *stochastic Euler scheme* for this SDE is componentwise

$$X_{n+1}^i = X_n^i + a^i(t_n, X_n)\, \Delta_n + \sum_{j=1}^{m} b^{i,j}(t_n, X_n) \Delta W_n^j \qquad (10.36)$$

for $i = 1, \ldots, n$, and the *Milstein scheme* is componentwise

$$X_{n+1}^i = X_n^i + a^i(t_n, X_n)\, \Delta_n + \sum_{j=1}^{m} b^{i,j}(t_n, X_n) \Delta W_n^j \qquad (10.37)$$

$$+ \sum_{j_1, j_2 = 1}^{m} L^{j_1} b^{i, j_2}(t_n, X_n) I_{(j_1, j_2) t_n, t_{n+1}}$$

for $i = 1, \ldots, n$, where the operators L^j are defined

$$L^j = \sum_{i=1}^{n} b^{i,j} \frac{\partial}{\partial x_i}$$

for $j = 1, \ldots, m$. As in the scalar case, the vector stochastic Euler and Milstein schemes here have strong convergence orders $\gamma = \frac{1}{2}$ and 1, respectively, under analogous conditions on the SDE coefficients.

MAPLE ASSISTANCE 95

Below we give MAPLE procedures Euler and Milstein that construct numerical schemes of the same names. Both procedures incorporate procedures L0 and LJ from Section 9.4.1 (recall that L^0 applied to x^i gives a^i L^j applied to x^i gives $b^{i,j}$), and hence these procedures need to be defined in the worksheet before Euler and Milstein can be applied. Both routines below are powerful enough to construct such strong schemes for vector Ito SDEs of any dimension.

```
> Euler:=proc(a::list(algebraic),b::list(list(algebraic)) )
> local i,u,soln;
> for i to nops(a) do
> soln[i]:=Y.i[n+1]=Y.i[n]+L0(x[i],a,b)*Delta[n]
> +sum('LJ(x[i],b,j)*Delta*W.j[n]'
> ,'j'=1..nops(op(1,b)));
> for u to nops(a) do
> soln[i]:=subs(x[u]=Y.u[n],soln[i]) od
> od;
> RETURN(eval(soln))
> end:
```

Note that the input parameters a and b must be of the same form as the input parameters a and b for procedures L0 and LJ. The syntax required for procedure Euler is typical of all the numerical procedures that we will subsequently encounter.

Procedure Milstein is next.

```
> Milstein :=
  proc(a::list(algebraic),b::list(list(algebrai c)))
> local u,i,soln;
> for i to nops(a) do
> soln[i]:=Y.i[n+1]=Y.i[n]+L0(x[i],a,b)*Delta[n]
> +sum('LJ(x[i],b,j)*Delta*W.j[n]','j'=1..nops(op(1,b)))
> +sum('sum('LJ(op(j2,op(i,b)),b,j1)*I[j1,j2]',
> 'j1'=1..nops(op(1,b)))','j2'=1..nops(op(1,b)));
> for u to nops(a) do
> soln[i]:=subs(x[u]=Y.u[n],soln[i])
> od;
> od;
> RETURN(eval(soln))
> end:
```

The syntax for Milstein is identical to that of Euler. The terms $I_{j1,j2}$ that appear in the output resulting from procedure Milstein represent the multiple integrals described earlier.

$$\cdots$$

A major difficulty in implementing the the Milstein scheme (10.37) is that the multiple stochastic integrals

$$I_{(j_1,j_2);t_n,t_{n+1}} = \int_{t_n}^{t_{n+1}} \int_{t_n}^{s_1} dW_{s_2}^{j_1} dW_{s_1}^{j_2}$$

for $j_1 \neq j_2$ cannot be expressed in terms of the increments $\Delta W_n^{j_1}$ and $\Delta W_n^{j_2}$ of the components of the Wiener processes as in the case $j_1 = j_2 = j$ where

$$I_{(j,j);t_n,t_{n+1}} = \frac{1}{2} \left\{ (\Delta W_n^j)^2 - \Delta_n \right\}.$$

The best strategy is to try to avoid having to use the multiple stochastic integrals $I_{(j_1,j_2);t_n,t_{n+1}}$ for $j_1 \neq j_2$, if at all possible. The coefficients of many SDEs encountered in applications often have a special structure, such as *additive noise* with

$$\frac{\partial}{\partial x_k} b^{i,j}(t, x) \equiv 0 \qquad (10.38)$$

for $i, k = 1, \ldots, n$ and $j = 1, \ldots, m$, or the *commutativity condition*

$$L^{j_1} b^{k,j_2}(t, x) \equiv L^{j_2} b^{k,j_1}(t, x) \qquad (10.39)$$

for $k = 1, \ldots, n$ and $j_1, j_2 = 1, \ldots, m$. For SDEs with additive noise, the difficult double stochastic integrals in the Milstein scheme (10.37) disappear, the scheme essentially reducing to the stochastic Euler scheme (10.36). For SDEs with commutative noise we can use the identities (see (9.46))

$$I_{(j_1,j_2);t_n,t_{n+1}} + I_{(j_2,j_1);t_n,t_{n+1}} = \Delta W_n^{j_1} \Delta W_n^{j_2} \qquad (10.40)$$

for $j_1, j_2 = 1, \ldots, m$ with $j_1 \neq j_2$. Inserting this into (3.4), we see that the *Milstein scheme for commutative noise* can be written as

$$X_{n+1}^i = X_n^i + a^i(t_n, X_n) \Delta_n + \sum_{j=1}^{m} b^{i,j}(t_n, X_n) \Delta W_n^j \qquad (10.41)$$

$$+ \frac{1}{2} \sum_{j_1=1}^{m} L^{j_1} b^{i,j_1}(t_n, X_n) \left\{ (\Delta W_n^{j_1})^2 - \Delta_n \right\}$$

$$+ \frac{1}{2} \sum_{\substack{j_1,j_2=1 \\ j_1 \neq j_2}}^{m} L^{j_1} b^{i,j_2}(t_n, X_n) \Delta W_n^{j_1} \Delta W_n^{j_2}.$$

MAPLE ASSISTANCE 96

Below a MAPLE procedure, called milcomm, is presented that checks for commutative noise.

```
>   milcomm:=proc()
>   local LJ1,LJ2,l,k,j1,j2,flag;
>   for k from 1 to nargs do
>   for j1 from 1 to nops(args[1]) do
>   for j2 from 1 to nops(args[1]) do
>   LJ1:=sum('op(j1,args[1])*diff(op(j2,args[k]),x[1])'
>   ,'l'=1..nargs);
```

```
>   LJ2:=sum('op(j2,args[1])*diff(op(j1,args[k]),x[1])'
>   ,'l'=1..nargs);
>   if LJ1<>LJ2 then
>   flag:=1;
>   fi;
>   od;
>   od;
>   od;
>   if flag=1 then
>   print('Commutative noise does not exist for this system');
>   else
>   print('This system has commutative noise');
>   fi;
>   end:
```

Input for milcomm should be entered as arrays separated by commas, where the ith array corresponds to the ith row of the diffusion matrix.

$$\cdots$$

However, we may have no choice because the SDE that we need to solve may not have commutative noise or some other convenient simplifying structure. Then, if we simply replace the $I_{(j_1,j_2);t_n,t_{n+1}}$ integrals in the Milstein scheme (10.37) by $\frac{1}{2}\Delta W_n^{j_1}\Delta W_n^{j_2}$ when $j_1 \neq j_2$, i.e use the scheme (10.41), we may only achieve the lower strong order of convergence $\gamma = \frac{1}{2}$.

Example 10.6.1 The 2-dimensional Ito SDE with a 2-dimensional Wiener process

$$dX_t^1 = dW_s^1, \qquad dX_t^2 = X_s^1 \, dW_s^2 \qquad (10.42)$$

does not have commutative noise since

$$L^1 b^{2,2}(t,x) \equiv 1 \neq 0 \equiv L^2 b^{2,1}(t,x).$$

The scheme (10.41) here with constant stepsize Δ is

$$X_{n+1}^1 = X_n^1 + \Delta W_n^1$$
$$X_{n+1}^2 = X_n^2 + X_n^1 \Delta W_n^2 + \frac{1}{2} \Delta W_n^1 \Delta W_n^2.$$

It can be shown for the initial value $X_0^1 = X_0^2 = 0$ that

$$\sqrt{\mathbb{E}\left(\left|X_{N_T}^2 - X_T^2\right|^2\right)} = \frac{1}{2} T^{1/2} \Delta^{1/2},$$

which means that the scheme converges only with strong order $\gamma = \frac{1}{2}$.

MAPLE ASSISTANCE 97

Below we use the previously defined procedure `milcomm` to illustrate how MAPLE can be used to show that the above SDE does not have commutative noise.

```
>  milcomm([1,0],[0,x[1]]);
```
   Commutative noise does not exist for this system

Next, a Milstein scheme is constructed for this vector SDE using procedure `Milstein`.

```
>  Milstein([0,0],[[1,0],[0,x[1]]]);
```

$$\text{table}([$$
$$1 = (Y1_{n+1} = Y1_n + \Delta W1_n)$$
$$2 = (Y2_{n+1} = Y2_n + Y1_n \Delta W2_n + I_{1,2})$$
$$])$$

$$\cdots$$

The preceding example does, however, provide a hint on how to simulate the double stochastic integral $I_{(1,2);t_n,t_{n+1}}$ for $j_1 \neq j_2$. Note that the solution of the SDE (10.42) with the initial value $X_0^1 = X_0^2 = 0$ is

$$X_t^1 = W_t^1, \qquad X_t^2 = I_{(1,2);0,t}.$$

We could apply the stochastic Euler scheme (10.36) to the SDE (10.42) over the interval $[t_n, t_{n+1}]$ with a smaller step size δ which is proportional to Δ_n^2, where $\Delta_n = t_{n+1} - t_n$ is the length of the time interval on which one wants to simulate $I_{(1,2);t_n,t_{n+1}}$. The initial conditions need to be appropriately reset before reapplying the stochastic Euler scheme on the next time interval.

10.7 MAPLE Session

M 10.1 *Consider equation:*

$$dX_t = \lambda X_t \, dt + \sigma \sqrt{X_t} \, dW_t, \tag{10.43}$$

which has been used as a square asset-price model. Here λ is the coefficient of the drift term and σ the coefficient of volatility. We find 300 paths of the solution of this equation for $\lambda = 0.05$, $\sigma = 0.9$, assuming the initial condition $X_0 = 100$. We will use the procedure `EulerScheme` defined on page 282 on the interval $0 \leq t \leq 1$ with stepsize $\Delta = dt = 0.01$. At every point $t_i = i \, dt$, $i = 0, \ldots, 100$ we will find the expectation and variance.

```
>  N := 100: npaths := 300: dt := 0.01:
>  for i from 1 to npaths do
   EulerScheme((t,x) ->0.05*x,(t,x)->.9*sqrt(x), 100,dt,N):
   p[i] := map(x->x[2],%)
   od:
   A := convert([seq(p[i],i=1..npaths)],matrix):
   for j from 1 to N+1 do
   m[j] := add(A[i,j], i = 1..npaths)/npaths:
   var[j] := add( (A[i,j] - m[j])^2, i = 1..npaths)/npaths:
   od:
```

Each path is represented by the row of matrix A. Note that we have $N + 1$ observations. On the other hand, each column represents observed values of the solution X_t at a fixed time t.

We plot the expectation with the 0.95-prediction interval for each observation. By $(1 - \alpha)$-prediction interval for a random variable X we mean an interval (a, b) such that $P(X \in (a, b)) = 1 - \alpha$.

```
>  p1 := plots[pointplot]([ seq([ (j-1)*dt,m[j] ],
   j=1..N+1) ]):
>  p2 := plots[pointplot]([ seq([ (j-1)*dt,m[j]
   +1.96*sqrt(var[j]) ], j = 1..N+1) ], symbol = POINT):
>  p3 := plots[pointplot]([ seq([ (j-1)*dt,m[j]
   -1.96*sqrt(var[j]) ], j = 1..N+1) ], symbol = POINT):
>  plots[display](p1,p2,p3);
```

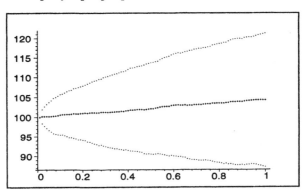

Actually, we have assumed that at each point $t = i\,dt$ the solution was normally distributed. Now, we check this graphically.

```
>  with(stats,statplots):
>  statplots[histogram]([seq(A[i,2], i = 1..npaths)],
   numbars = floor(log[2](npaths)), area = count);
```

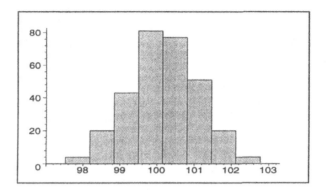

```
>  statplots[histogram]([seq(A[i,N+1], i = 1..npaths)],
   numbars = floor(log[2](npaths)), area = count);
```

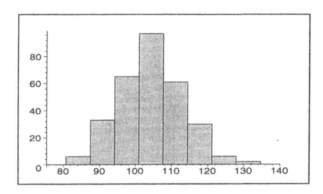

We can look at all histograms forming ing the animation.

```
>  plots[display](
   [seq( statplots[histogram](
   [seq(A[i,k], i = 1..npaths)],
   numbars = floor(log[2](npaths)), area = count),
   k =2..N+1)],
   insequence=true,view = [80..140,0..100]
   ):
```

The normality of X_t appears to be questionable. In fact, it fails for some
other parameters.

M 10.2 *We will determine and plot confidence intervals for the mean of the
solution in the above example. We will use approach discussed on the page
283.*

```
>  L := 10: K := 30:
>  B := matrix(L,N+1,[]);
```
$$B := \text{array}(1..10,\ 1..100,\ [])$$

```
>  for j from 1 to N+1 do
   for i0 from 1 to L do
   B[i0,j] := add(A[i,j], i = 1+K*(i0-1)..K*i0)/30:
   od:
   od:
>  for j from 1 to N+1 do
   mB[j] := add(B[i,j], i = 1..L)/L:
   varB[j] := add( (B[i,j] - mB[j])^2, i = 1..L)/(L-1):
   od:
>  alpha := 0.05:
>  t := stats[statevalf,icdf,studentst[L-1]](1 - alpha/2);
```
$$t := 2.262157163$$
```
>  p1 := plots[pointplot]([ seq([ (j-1)*dt,m[j] ],
   j=1..N+1) ]):
>  p2 := plots[pointplot]([ seq([ (j-1)*dt,m[j]
   +t*sqrt(varB[j]/L) ], j = 1..N+1) ], symbol = POINT):
>  p3 := plots[pointplot]([ seq([ (j-1)*dt,m[j]
   -t*sqrt(varB[j]/L) ], j = 1..N+1) ], symbol = POINT):
```

We display the results using a similar scale as it was for the prediction intervals above.

```
>  plots[display]([p1,p2,p3],view = [0..1,90..120]);
```

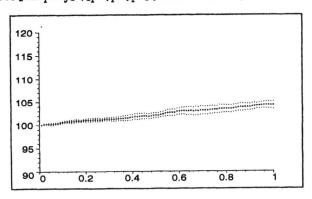

M 10.3 *Construct a Milstein scheme for the following system, which is a model called the Bonhoeffer-Van der Pol oscillator. These equations can be used to model the firing of a single neuron.*

$$dX_t^1 = c\{X_t^1 + X_t^2 - \frac{1}{3}(X_t^1)^3 + z\}dt + \sigma dW_t$$

$$dX_t^2 = -\frac{1}{c}\{X_t^1 + bX_t^2 - a\}dt \tag{10.44}$$

where $\sigma > 0$ is a noise intensity parameter, X_t^1 is the negative of the membrane voltage, X_t^2 is the membrane permeability and a, b, c, z are external parameters.

Below, we use the previously defined procedure Milstein to construct the appropriate Milstein scheme for the above system.

```
>   Milstein([c*(x[1]+x[2]-1/3*(x[1])^3+z),
    -1/c*(x[1]+b*x[2]-a)],[[sigma],[0]]);
```

table([

$$1 = (Y1_{n+1} = Y1_n + c\,(Y1_n + Y2_n - \frac{1}{3}\,Y1_n{}^3 + z)\,\Delta_n + \sigma\,\Delta W1_n)$$

$$2 = (Y2_{n+1} = Y2_n - \frac{(Y1_n + b\,Y2_n - a)\,\Delta_n}{c})$$

])

Note that since the noise is additive here, the Milstein scheme above will be identical to the corresponding Euler scheme.

M 10.4 *Write a* MAPLE *procedure in the same style as procedures* Euler *and* Milstein *that will construct the weak scheme of order 2*

$$Y_{n+1} = Y_n + a\Delta_n + b\Delta_n W_n + \frac{1}{2}\{aa' + \frac{1}{2}b^2 a''\}\Delta_n^2$$

$$+ \{ab' + \frac{1}{2}b^2 b''\}(\Delta_n \Delta W_n - \Delta Z_n) + ba'\Delta Z_n$$

$$+ \frac{1}{2}bb'((\Delta W_n)^2 - \Delta_n) \tag{10.45}$$

for an autonomous scalar Ito SDE $dX_t = a(X_t)dt + b(X_t)dW_t$, where $a = a(x)$, $b = b(x)$ and the $'$ denotes differentiation with respect to x. Then use the procedure to construct a weak scheme of order 2 for the scalar Ito SDE $dX_t = 4\alpha^2 X_t^3 dt + 2\alpha X_t^2 dW_t$.

Below we give a suitable MAPLE procedure, called Weak2. Note that since this procedure is designed for scalar Ito SDEs, it uses the operator procedures L0 and L1 as defined in section 8.4. The application of these operator combinations, as given in the procedure below, results in the various coefficients in the above scheme.

```
>   Weak2:=proc(a,b)
>   Y[n+1]=Y[n]+L0(x,a,b)*Delta[n]+L1(x,b)*Delta*W[n]
>   +L0(L0(x,a,b),a,b)
>   *1/2*Delta^2+L0(L1(x,b),a,b)*(Delta*W[n]*Delta[n]
>   -Delta*Z[n])
```

```
>  +L1(L0(x,a,b),b)*Delta*Z[n]+L1(L1(x,b),b)*1/2
>  *(Delta*W[n]^2-Delta[n]);
>  subs(x=Y[n],%);
>  end:
```

The input parameters here are the drift a and diffusion b of the scalar Ito SDE. Input must be given in terms of x. Below we apply this procedure to the given Ito SDE.

```
>  Weak2(4*alpha^2*x^3,2*alpha*x^2);
```

$$Y_{n+1} = Y_n + 4\alpha^2 Y_n^3 \Delta_n + 2\alpha Y_n^2 \Delta W_n + 48\alpha^4 Y_n^5 \Delta^2$$
$$+ 24\alpha^3 Y_n^4 (\Delta W_n \Delta_n - \Delta Z_n) + 24\alpha^3 Y_n^4 \Delta Z_n$$
$$+ 4\alpha^2 Y_n^3 (\Delta W_n^2 - \Delta_n)$$

Since this is a weak scheme, we can replace the random variables ΔW_n and ΔZ_n with other random variables which are easier to generate. Note also that the scalar Ito SDE here is reducible and hence possesses a known explicit solution.

M 10.5 *Apply procedure* Taylor1hlf *from the* MAPLE *software package called* stochastic *to the following system of Ito SDEs to construct an appropriate strong scheme of order 1.5. (This system is used in radio astronomy).*

$$dY_t^1 = a\cos(b + Z_t)dt + rdW_t^1$$
$$dY_t^2 = a\sin(b + Z_t)dt + rdW_t^2 \qquad (10.46)$$
$$dZ_t = -\beta Z_t dt + \sigma\sqrt{2\beta}dW_t^3$$

The procedures Taylor1hlf can be found in the MAPLE software package called stochastic, which also contains many other routines that are useful in the area of stochastic numerics. The procedures Weak2 in the previous examples, together with procedures Euler and Milstein defined earlier come also from this package or its earlier versions. Download the stochastic package from the Internet web address

http://www.math.uni-frankfurt.de/~numerik/kloeden

After downloading the stochastic package, we can see the various routines that it contains using the with command.

```
>  with(stochastic);
```

[*Euler, L0, LFP, LJ, MLJ, Milstein, SL0, Taylor1hlf, Taylor2, ap,
bpb, chainrule, colour, comm1, comm2, conv, correct, explicit,
itoformula, linearize, linearsde, milcomm, momenteqn, pa,
pmatrix2pvector, position, pvector2pmatrix, reducible, sphere,
wkeuler, wktay2, wktay3*]

Next, we apply procedure `Taylor1h1f` to construct the appropriate scheme.

```
>   Taylor1h1f([a*cos(b+x[3]),a*sin(b+x[3]),-beta*x[3]],
>   [[r,0,0],[0,r,0],[0,0,sigma*sqrt(2*beta)]]);
```

$$Y1_{n+1} = Y1_n + a\cos(b + Y3_n)\,\Delta_n$$
$$+ \frac{1}{2}\left(\beta\,Y3_n\,a\sin(b + Y3_n) - \sigma^2\,\beta\,a\cos(b + Y3_n)\right)\Delta_n^2 + r\,\Delta W1_n$$
$$- \sigma\sqrt{2}\sqrt{\beta}\,a\sin(b + Y3_n)\,I_{3,0}$$

In the above scheme, the MAPLE variables $Y1_n$, $Y2_n$ and $Y3_n$ denote approximations to the variables Y_t^1, Y_t^2 and Z_t, respectively, in equation (10.46).

10.8 Exercises

E 10.1 Rewrite the `Eulerscheme` procedure using two-point distribution instead of the normal distribution. Then use it for determining confidence interval of the solution mean for the equation (10.43). Compare the times needed for producing 300 solution paths, using 200 steps. You may want to use the MAPLE procedure **time** here.

E 10.2 Show that the stochastic Euler scheme is <u>exact</u> for the stochastic differential equation

$$dX_t = \alpha\,dt + \beta\,dW_t,$$

i.e. where α and β are constants, has the orders of strong and weak convergence are $\gamma = \beta = \infty$.

E 10.3 Find $(1 - \alpha)$-prediction intervals for the solution X_t of the equation (10.43) of the form $\left(q_{\frac{\alpha}{2}}, q_{1-\frac{\alpha}{2}}\right)$, where q_s is the sample s-th-quantile, i.e. a value that separates the sample into 2 parts. One having the portion s of the sorted sample which is less than the quantile value, the other have the $1 - s$ portion of the sample which is greater than the quantile value.

E 10.4 Prove that the random variable ΔZ_n define in (10.21) has expectation, variance and correlation with ΔW_n given by (10.22). Then write a MAPLE procedure to generate the random variables ΔW_n and ΔZ_n.

E 10.5 Write a MAPLE procedure to generate the three-point random variable (10.34).

E 10.6 Show that the scalar Ito SDE

$$dX_t = aX_t\,dt + b\,dW_t^1 + c\,dW_t^2,$$

where W_t^1 and W_t^2 are independent Wiener processes and a, b, c are constants, has commutative noise. Derive the Milstein scheme for this SDE and simplify it using the relationship (10.24) between multiple stochastic integrals. Compare the result with the Milstein scheme for commutative noise obtained by applying the MAPLE procedure **milcomm**.

E 10.7 Show that the 2-dimensional Ito SDE (10.42) with the initial condition $X_{t_n}^1 = W_{t_n}^1$ and $X_{t_n}^2 = 0$ has the solution $X_t^1 = W_t^1$ and $X_t^2 = I_{(1,2);t_n,t}$ for $t \geq t_n$.

Bibliographical Notes

This book covers a considerably large amount of mathematics, in particular probability theory and statistics, measure and integration, stochastic analysis, stochastic differential equations, and numerical analysis. The authors are well aware that some topics have been addressed only very briefly here, so would like to provide some suggestions for readers who wish to fill in on background topics and to broaden their knowledge and expertise.

There are many nice textbooks on probability theory. We recommend those of Ash [2] and Billingsley [3] for the reader who is interested in a more theoretical approach. Halmos' book [13] is a classical textbook of measure and integration theory, that is still very useful. On the other hand, an applications oriented reader may prefer to read Kalbfleisch's book [16], volume 1. In addition, Feller's classical book [9] is a mine of very interesting examples and ideas for those who want to learn a lot more about probability theory and its applications. Also recommended are the books [7], [15], [21], [28], [32], [33].

The reader who does not have a strong mathematical background, but wants to study statistics more systematically may find the books by Kalbfleisch [16], volume 2, and Aczel [1] useful, while the mathematically more advanced reader will find the book by Silvey [31] very interesting. The books by Lehmann [22], [23] provide a presentation of statistical inference for the mathematically sophisticated. New statistical methods that exploit the power of modern computers are addressed, for example, in [6] and [30].

An exhaustive treatment of Markov chains can be found in [14], while the textbooks [3], [7], [9] mentioned above each contain chapters on the subject. Stochastic processes and stochastic analysis are treated in depth in [5], [18] and from the perspective of stochastic differential equations in [11] and [25], as well as in [19] where numerical methods for stochastic differential equations are presented in a systematic way. More advanced topics on numerical methods for stochastic differential equations are considered in Milstein's book [24], while computer experiments with PASCAL software can be found in [20]. More advanced treatment of the subject in the context of partial differential equations can be found in [12].

At present there are not many books on probability and statistics using Maple. One is the book of Karian and Tanis [17], who have developed a Maple package containing a lot of very useful statistical procedures.

References

[1] A. Aczel, *Complete Business Statistics*, Irving–McGraw–Hill, Boston, 1999.

[2] R. Ash, *Real Analysis and Probability*, Academic Press, New York, 1972.

[3] P. Billingsley, *Probability and Measure*, John Wiley & Sons, New York, 1979.

[4] R.L. Burden and J.D. Faires, *Numerical Analysis*, Fourth Edition, PWS-Kent Publishing Co., Boston, 1989.

[5] Z. Brzeźniak, T. Zastawniak, *Basic Stochastic Processes*, Springer, Heidelberg, 1999.

[6] B. Efron, *An Introduction to the Bootstrap*, Chapman and Hall, New York, 1993.

[7] K.L. Chung, *Elementary Probability Theory with Stochastic Processes*, Springer-Verlag, Heidelberg, 1974.

[8] S. Cyganowski, L. Grüne and P.E. Kloeden, MAPLE *for Stochastic Differential Equations*, in *Theory and Numerics of Differential Equations, Durham 2000*, (Editors: A. Craig and J. Blowey), Springer-Verlag, Heidelberg, 2001.

[9] W. Feller, *An Introduction to Probability Theory and Its Applications*, John Wiley & Sons, New York, 1961.

[10] W. Gander and J. Hrebicek, *Solving Problems in Scientific Computing using Maple and Matlab*, Second Edition, Springer-Verlag, Heidelberg, 1995.

[11] T.C. Gard, *Introduction to Stochastic Differential Equations*, Marcel Dekker, New York, 1988.

[12] C. Graham, et al, *Probabilistic Models for Nonlinear Partial Differential Equations*, Lecture Notes in Mathematics, 1627, Springer-Verlag, 1996.

[13] P.R. Halmos, *Measure Theory*, Reprint, Springer-Verlag, Heidelberg, 1974.

[14] M. Iosifescu, *Finite Markov Processes and Their Applications*, John Wiley & Sons, New York, 1980.

[15] 1.Jacod and P. Protter, *Probability Essentials*, Springer-Verlag, Berlin, 2000.

[16] J.G. Kalbfleisch, *Probability and Statistical Inference I and II*, Springer-Verlag, Heidelberg, 1979.

[17] Z.A. Karian and E.A. Tanis, *Probability and Statistics, Explorations with Maple*, Prentice Hall, Edgewood Cliffs, 1999.

[18] I. Karatzas and S.E. Shreve *Brownian Motion and Stochastic Calculus*, Springer-Verlag, Heidelberg, 1988

[19] P.E. Kloeden and E. Platen, *Numerical Solution of Stochastic Differential Equations*, Springer-Verlag, Heidelberg, Third printing, 1999

[20] P.E. Kloeden, E. Platen and H. Schurz, *Numerical Solution of Stochastic Differential Equations through Computer Experiments*, Springer-Verlag, Heidelberg, Revised reprinting, 1997.

[21] *Lexicon der Stochastik*, Academie-Verlag, Berlin 1983.

[22] E.L. Lehmann, *Theory of Point Estimation,* John Wiley & Sons, New York, 1983.

[23] E.L. Lehmann, *Testing Statistical Hypothesis,* John Wiley & Sons, New York, 1986.

[24] G.N. Milstein, *The Numerical Integration of Stochastic Differential Equations,* Kluwer Academic Publishers, Dordrecht, 1995.

[25] B. Oksendal, *Stochastic Differential Equations,* Springer-Verlag, Heidelberg, 1985.

[26] J. Ombach, *Introduction to the Calculus of Probability,* Wydawnictwo IM AGH, Kraków, 1997, in Polish.

[27] J. Ombach, *Computer Assisted Calculus of Probability – Maple,* Wydawnictwo Uniwersytetu Jagiellońskiego, Kraków, 2000, in Polish.

[28] J. Pitman, *Probability,* Springer-Verlag, Heidelberg, 1999.

[29] H.L. Royden, *Real Analysis,* Third Edition, Macmillan, London, 1989.

[30] B.W. Silverman, *Density Estimation for Statistics and Data–Analysis,* Chapman and Hall, London, 1988.

[31] S.D. Silvey, *Statistical Inference,* Halsted, New York, 1975.

[32] J.L. Snell, *Introduction to Probability,* Random House, New York, 1988.

[33] D.R. Stirzaker, *Probability and Random Variables: A Beginners Guide,* Oxford Univ. Press, Oxford, 1999.

Index

Universitext

Aksoy, A.; Khamsi, M. A.: Methods in Fixed Point Theory

Alevras, D.; Padberg M. W.: Linear Optimization and Extensions

Anderson, M.: Topics in Complex Analysis

Aoki, M.: State Space Modeling of Time Series

Aupetit, B.: A Primer on Spectral Theory

Bachem, A.; Kern, W.: Linear Programming Duality

Bachmann, G.; Narici, L.; Beckenstein, E.: Fourier and Wavelet Analysis

Badescu, L.: Algebraic Surfaces

Balakrishnan, R.; Ranganathan, K.: A Textbook of Graph Theory

Bapat, R.B.: Linear Algebra and Linear Models

Balser, W.: Formal Power Series and Linear Systems of Meromorphic Ordinary Differential Equations

Benedetti, R.; Petronio, C.: Lectures on Hyperbolic Geometry

Berberian, S. K.: Fundamentals of Real Analysis

Berger, M.: Geometry I, and II

Bliedtner, J.; Hansen, W.: Potential Theory

Blowey, J. F.; Coleman, J. P.; Craig, A. W. (Eds.): Theory and Numerics of Differential Equations

Börger, E.; Grädel, E.; Gurevich, Y.: The Classical Decision Problem

Böttcher, A; Silbermann, B.: Introduction to Large Truncated Toeplitz Matrices

Boltyanski, V.; Martini, H.; Soltan, P. S.: Excursion into Combinatorial Geometry

Boltyanskii, V. G.; Efremovich, V. A.: Intuitive Combinatorial Topology

Booss, B.; Bleecker, D. D.: Topology and Analysis

Borkar, V. S.: Probability Theory

Carleson, L.; Gamelin, T.: Complex Dynamics

Cecil, T. E.: Lie Sphere Geometry: With Applications of Submanifolds

Chae, S. B.: Lebesgue Integration

Chandrasekharan, K.: Classical Fourier Transform

Charlap, L. S.: Bieberbach Groups and Flat Manifolds

Chern, S.: Complex Manifolds without Potential Theory

Chorin, A. J.; Marsden, J. E.: Mathematical Introduction to Fluid Mechanics

Cohn, H.: A Classical Invitation to Algebraic Numbers and Class Fields

Curtis, M. L.: Abstract Linear Algebra

Curtis, M. L.: Matrix Groups

Dalen, D. van: Logic and Structure

Das, A.: The Special Theory of Relativity: A Mathematical Exposition

Debarre, O.: Higher-Dimensional Algebraic Geometry

Demazure, M.: Bifurcations and Catastrophes

Devlin, K. J.: Fundamentals of Contemporary Set Theory

DiBenedetto, E.: Degenerate Parabolic Equations

Diener, F.; Diener, M.: Nonstandard Analysis in Practice

Dimca, A.: Singularities and Topology of Hypersurfaces

DoCarmo, M. P.: Differential Forms and Applications

Duistermaat, J. J.; Kolk, J. A. C.: Lie Groups

Edwards, R. E.: A Formal Background to Higher Mathematics Ia, and Ib

Edwards, R. E.: A Formal Background to Higher Mathematics IIa, and IIb

Emery, M.: Stochastic Calculus in Manifolds

Endler, O.: Valuation Theory

Erez, B.: Galois Modules in Arithmetic